WHY IS A FLY NOT A HORSE?

WHY IS A FLY NOT A HORSE?

DIMENTICARE DARWIN

GIUSEPPE SERMONTI

Description

This book is an English translation of *Dimenticare Darwin* ("Forget Darwin"), which was published in Milan by Rusconi in 1999. The title for this translation comes from Chapter VI.

Giuseppe Sermonti is a retired Professor of Genetics at the University of Perugia and the chief editor of *Rivista di Biologia/Biology Forum*, one of the oldest still-published biology journals in the world. He is the author of seven books, including *Genetics of Antibiotic-producing Microorganisms* (1969), *Dopo Darwin* (1980), *Fiabe dei fiori* (1992), and *Il mito della grande madre* (2002).

Copyright Notice

Publisher's Note

This book is part of a series published by the Center for Science & Culture at Discovery Institute in Seattle. Previous books in that series include *Are We Spiritual Machines?: Ray Kurzweil vs. The Critics of Strong A.I.* by Jay W. Richards et. al. and *Getting the Facts Straight: A Viewer's Guide to PBS's Evolution* by the Discovery Institute.

Library Cataloging Data

Why is a Fly Not a Horse? by Giuseppe Sermonti (1925–)
Translated by Brendan White and edited by Jonathan Wells.
173 pages, 6 x 9 x 0.4 inches & 0.6 lb, 229 x 152 x 10 mm. & 0.26 kg
Subject: Evolution (Biology)—Philosophy
Subject: Natural Selection—Philosophy
Library of Congress Control Number: 2005904896
QH360.5 .S46 2005 and 576.8/01 21
ISBN-10: 0-9638654-7-1   ISBN-13: 978-0-9638654-7-2 (paperback)
ISBN-10: 0-9638654-8-X   ISBN-13: 978-0-9638654-8-9 (hardback)

Publisher Information

Discovery Institute Press, Discovery Institute, 1511 Third Avenue, Suite 808, Seattle, WA 98101. Internet: http://www. discovery.org/

Published in the United States of America on acid-free paper.
First Edition, First Printing. June 2005.

DEDICATION

To my eight grandchildren.

CONTENTS

ACKNOWLEDGEMENTS

This part of a book is usually devoted to expressions of gratitude, which are often acknowledgements of debts and an exchange of courtesies that have no interest for the reader. Instead, I will render an account of ventures, meetings and friendships to which I am indebted for this book.

Following my book *Dopo Darwin* with R. Fondi in 1980, which against my nature made me into a knight errant, and after heated post-publication exchanges that taught me nothing, I embarked on two activities I had not previously envisaged that changed my way of thinking as a scientist quite apart from my revolt against various stupidities, conspiracies and conformities. In 1979, I was chosen by Aldo Spirito to edit *Rivista di Biologia* (founded in 1919). To be an editor is a gratifying profession. It brings one into contact with other people's thoughts, gives one an opportunity to discuss those thoughts, and opens up an impartial intellectual correspondence free of the obligation to defend positions.

Among the contributors to *Rivista,* I made the acquaintance of the great Swedish geneticist Søren Løvtrup, who introduced me to the field of epigenetics, and the genial Japanese lepidopterist Atuhiro Sibatani, later to become Rector of Kyoto University, who revealed to me the biological world of the East. Løvtrup and Sibatani became deputy editors and my counselors. The former introduced me to Gerry Webster and Brian Goodwin, who collaborated with *Rivista,* and whose book on biological form I edited for Armando Armando (1988). In 1986, Sibatani organized an international Congress on Structuralism in Biology at Osaka, which brought together some of the brightest dissenters from neo-Darwinism and enabled me to meet them.

At Osaka I met the Australian systematist Hugh Paterson, the French mathematician René Thom (who remained my friend), the extraordinary Anglo-Chinese lady Mae-Wan Ho, the New Zealander David Lambert, the Japanese Susumo Ohno and the Italian Franco Scudo from Pavia. These people, with their ideas, their work and their different viewpoints, have enriched me with a fund of knowledge that permeates this entire book. Sibatani taught me the arbitrary nature of the genetic code; Webster and Goodwin introduced me to the generative paradigm and the morphogenetic field. From Paterson I learned the concept of species as systems of recognition; from Thom I picked up the idea that form draws the species unto itself. Mae-Wan Ho entertained me with coherent systems and liquid crystals; Ohno introduced me to the music of DNA; Lambert helped me to explode the myth of the peppered moth; and Scudo taught me to distinguish between Darwin and his latter-day followers.

Those who attended the congress in Japan formed the Osaka Group for the Study of Dynamic Structure in 1987, and I, with *Rivista,* took up editorship of its publications. The Group subsequently met in Prague, Cornwall ("Gaia"), Moscow, Mexico, Potsdam and Frankfurt. In Cornwall I met Teddy Goldsmith, from whom I learned to regard ecology as a moral problem. In Moscow I met Lev Beloussov, representative of the great Russian morphological school, which the USSR persecuted and the West ignored. Since 1991, the Italian section of the Osaka Group has met every two years (in Genoa, Turin, Siena, Milan and Perugia), enabling me to preside at meetings of the most vital group of Italian biologists—including some who were initially my opponents. I also wish to acknowledge the biophysicist Renzo Morchio, the botanist Silvano Scannerini, the physicists Giuliano Preparata and Emilio Del Giudice, and the zoologist Michele Sarà.

The fraternal support of Ted Goldsmith and an unexpected encounter with the people of the Discovery Institute paved the way for the English translation of this book, which was done by Brendan White. I feel particularly grateful to Jonathan Wells for his patient and shrewd editing of the final text, for which I take complete responsibility. I apologize to my old friends for possible misinterpretations or omissions in referring to their ideas in this present venture.

A special word of gratitude goes to my wife Isabella Spada—with whom I talked at length as this book developed—and to my assistant Isabella Borghesi of Perugia, who helped me in the final editing of the manuscript. Anna Petris reviewed some chapters.

"EVOLUTION IS DEAD"

In the spring of 1982 I was invited to the Pius IV Pavilion in the Vatican gardens for a meeting of the Pontifical Academy of Sciences. The subject was the evolution of primates. At the close of the meeting everyone felt that some conclusion should be drawn, and the South African paleontologist Phillip Tobias proposed a motion to the effect that the origin of man and the primates was covered by the theory of evolution. Jérôme Léjeune, the French cytologist, objected: "There is no theory of evolution!" No one disagreed, and the meeting concluded with a Solomonic pronouncement that the evidence renders the application of the concept of divergent descent to man and other primates beyond dispute.

There never really has been a scientific "theory" of evolution. "Divergent descent" is an evasive way of stating ancient relationships among living beings, but evolution is about the ways and the mechanisms whereby species emerged and differentiated, say, from the amoeba to the elephant, from bacterium to man or, as the current fashion has it, from molecules to man. Definitions of the concept of evolution have to be looked up, even today, in dictionaries. Webster has: "A process of continuous change from a lower, simpler, or worse condition to a higher, more complex, or better state." Other lexicographers typically speak of an unremitting process, based on diverse factors, that slowly and gradually changes living organisms from lower, rudimentary forms to increasingly more complex forms.

Evolutionism is really more of a paradigm or methodology than a theory. For its present-day supporters, the important thing about evolution is that it was due to natural causes. That the process has been continuous, unremitting, and gradual, and improving organisms all the time, is of little interest to the supporters of the theory. Some even doubt whether such characteristics are real, while others consider them irrelevant.

As a matter of fact, all the currently envisioned physical causes of evolution are either degenerative or conservative; therefore not one of them guarantees passage from the simple to the complex, from the inferior to the better. Among them there is a vague promise of progress, but in a tautological and clumsy sense; and there is concession to gradualism, though empirical observation seems to exclude this, or at least fails to demonstrate it.

The underlying viewpoint that the adoption of materialism has imported into biology is the second law of thermodynamics, the principle of entropy. According to this law, every closed—i.e., isolated—system tends to uniformity, to a leveling, like a sand castle that time wears down and no future can ever build up again. Another way of stating the second law is this: A system in isolation does not pass twice through the same state. In fact, the term "entropy" means, precisely, "evolution." It affirms that the physical world tends to disorder—exactly the opposite of the way biologists try to use "evolution."

Jacques Monod (1971) believes that entropy enters into biological evolution in the form of irreversibility: Evolution does not retrace its own steps. Although entropy precludes complexity and disintegrates sand castles, Monod evades the problem with a sophism. The principle of entropy, he argues, is a statistical law, so it does not necessarily preclude the possibility that a macroscopic system, for the briefest of periods and for a moment of negligible duration, might climb a tiny bit back up the slope of entropy. If nature had at its disposal some mechanism that could capture and immobilize these rare, elusive, upward movements, it might be able to construct the complex from the simple, order from disorder, the sand castles from the windswept level beach.

For Monod, there is such a mechanism: cumulative natural selection. But if we are to build a sand castle (and a bacterium is vastly more complex than a sand castle) by capturing grains of sand tossed hither and thither by a storm, we would need to have the plan of the castle beforehand. Only then could we discern which rare and fortunate movements of the sand were correct. We would also need some system to protect the emerging edifice from its tendency to fall apart.

Other authorities have defended the plausibility of life's upward climb by maintaining that the living thing is not system in isolation, and that drawing energy from its environment enables it to defy the principle of entropy. This could have been argued by Darwin, but it cannot be consistently maintained by latter-day neo-Darwinists who affirm the so-called Central Dogma of molecular biology. According to that principle, the depository of heredity, DNA, remains in absolute isolation. DNA goes on its way indifferent to the pressures of the environment or the blandishments of the organism.

Natural selection could perhaps be invoked as a mechanism accounting for the survival of species. But the claim that natural selection is creative of life, of life's essence and types and orders, can only leave us dumbstruck. Natural selection only eliminates, and its adoption as a mechanism of origin is like explaining "appearance" by "disappearance." Many times a year a language dies; does this explain the origin of languages? It is to be hoped that the third millennium will look back on the twentieth century's wild guesses about evolution as a Big Joke in which the illustrious founders of molecular biology were able to indulge. The neo-Darwinian theory embraced by the founders of molecular biology is rather like saying that the text of *The Iliad* came into being by mere chance, one step at time, letter by letter, emerging out of some lower "organism" a few verses long.

Natural selection (which should be more accurately termed "differential survival") is not in doubt. No one has ever denied it. Without going into the subject at length, I will say that it chiefly eliminates the abnormal, the marginal, the out-of-bounds, and keeps natural populations within the norm. But this is clearly a conservative, policing role. There has been authoritative talk of Stabilizing Selection, as being a process for protecting a species from deviations: Might it be that, when environmental conditions change, selection can privilege those best adapted to the new situation? For a species to set aside its own founding values only produces suffering; and no one has yet shown that institutionalized suffering is the way to innovate life. Nowadays there is widespread enthusiasm for the idea that the principal mechanism in the origin of species is isolation, whether geographic, ecological or reproductive; but here selection has a modest role, if it has any role at all.

In this context, the scrambling of traits made possible by sexuality is the conservative process par excellence. It is the antithesis of isolation, and promotes continual mixing of the genetic endowments of the offspring of a population. It ferrets out every marginal deviation and mixes it again within the norm, or else rejects it. And yet, for some time sexuality was seen as the elixir of variation by reason of its combining abilities, which—so it was thought—would allow a limited number of variations to associate and form the maximum number of innovations and produce even Olympic champions. This way of seeing things was bound up with the idea that every gene corresponded to a character, and combining genes would have the effect of matching up characters. But the relationship between genes and characters has proved to be ever more complex and obscure; and mutations are not boutonnieres. The age-old precept retains its validity, therefore: If you want to isolate things it is better not to mix them together.

Mutations seemed so certainly the primary source of evolution that life was defined as Reproduction with Mutation. Actually, in the wild, morphological mutation is the absolute rarity; and when academic textbooks come to the chapter on mutations they give as illustrations the usual short-legged Ancon sheep, the albino baby, the stunted plant. But these, you may be sure, are assisted mutations, because nature would never have tolerated them. From the molecular standpoint, i.e. variations in the DNA text, mutation is a degenerative phenomenon, a copying error, a product of entropy in the genetic endowment. Redundancies within the genetic code and the interchangability of many amino acids render many mutations "neutral" vis-à-vis the phenotype. The cell possesses defense mechanisms (such as repair enzymes and "chaperones") to protect it from mutation, and populations of organisms can call on processes (such as selection and sexuality) that eliminate mutants. Were it not for these defenses, mutations would in no time destroy all genetic texts. Their effect in all instances is to demolish. To say that blind mutations are the driving principle of the world, and to rely on the rare fortunate mistake, is a poor resource, quite apart from the fact that transgressions of the kind needed by Darwinian evolution have never been documented.

Now that I have made it clear, as far as I was able, that the molecular mechanisms invoked to explain evolution are all fundamentally either degenerative or conservative, I want to stress another point: The modern molecular theory of evolution is stitched together from mechanisms—mechanisms that take precious little account of the facts, or of the forms of life and their history. The molecular revolution consisted precisely in disregarding natural observations and in disdaining forms. Given that everything is supposed to happen by chance, so evolution could have followed any path, via organisms utterly different from those with which we are acquainted, the real story of the emergence of life and its journey toward complexity are not of the slightest interest.

To the question "Were Life to be born again, would we have the same species?" the avant-garde evolutionist replies: "Indeed not!" If this is the case, what interest can he have in fossils, faunas and zoos?

Molecular biology has cultivated a passion for abstract "life," for possible lives, for models of life—models that take offense at any particular form, because a form claims the right to be what it is; whereas, according to neo-Darwinism, a form has no such right, because it could have been that which it is not—it might not have existed at all. Molecular evolution is the theory according to which lions, flies, cypresses, daisies, and (of course) humans might never have existed, in fact should not have existed. Do not ask how it is that nevertheless these forms are there. This is an ancient question, a question put by the nineteenth-century naturalists. Today the question has changed. It has shifted to mechanisms and their use. The highest aspiration that a molecular biologist can have is to be able to concoct life in a test tube, or at least to modify it as he pleases, to become a "genetic designer" rather than a contemplator of the beauty of the world.

An issue of *New Scientist* in October 1998 bore the title on the cover "Evolution is Dead." Yet the relevant article doesn't even mention evolution—only the genetic engineering of the human germ line. In the article, a certain Dr. G. Stock declares: "Technology will take the place of evolution, and the time scale will be much faster. Men are becoming objects of a conscious project." In other words, once we are able to create superman in a test tube, won't this furnish sufficient proof that a species can be formed using natural and mechanical instruments alone? Is this not sufficient to banish any idea of finality, vitalism and the Great Watchmaker? Let us omit the mysterious and the unverifiable journey of nature that in three billion years moved from non-life and led to the first throbbings of tiny marine animals; let us forget the mysterious histories that in hundreds of millions of years generated the forms of mammals; let us shelve the improbable careers through which a recent ape transformed itself into a biped. In other words, forget evolution and transfer the glory to the species concocted in the laboratory, to the Faustian superman, to a process safer, more expeditious and more exclusive to Science. The rough, eternal labor of a wasteful—and what's more, blind—demiurge no longer has anything interesting to teach. The biotechnologists celebrate the obsequies of that lesser god who did himself the honor of making himself in our image and likeness—and, while they are at it, they celebrate the obsequies of evolution in the process.

"Poor, pitiful evolution," the *New Scientist* insisted in September 2000. In billions of years it has limited itself to building up "proteins constructed from just twenty amino acids and DNA stitched together from an alphabet of four measly chemical letters." One "can't help but lament what might have been!"

Having shed his ancient Lord and Master with its clumsy Creation, the molecular biologist has also forfeited that pride of species that, in the traditional representation, placed *Homo* at the top of the tree. If ants were to take over control of the Earth, the last biologist alive in the moment of his own dissolution would say, "All to the good—it shows that the ants were better than we."

If only to touch on historical facts documented by fossil reconstructions, we can say that these, if anything, demonstrate events that differ markedly from the dictionary definitions of evolution—events that in any case neither contradict nor confirm evolutionary statements, since these can be neither denied nor verified. Single-celled life appeared four billion years ago, when the Earth itself had only just formed; and one can assume that, with microbes, evolution in the direction of complexity had accomplished ninety percent of its journey. Multicellular organisms emerged over a brief period, in all their "types," half a billion years ago without intermediate or premonitory forms. Modern mammals also exploded, quite distinct in their "orders," practically all together, at the beginning of the Paleocene. The supposed continuous change, the "slow, gradual passage," and even the process of improvement in the direction of complexity are anything but documented. But current theory refuses to explain particular facts about life because its major premise is that these facts have no sense, order or "intention;" and that, whatever they are, they are indifferent....

Here, perhaps, is the only claim to be factual, when we are told that organisms are all "useful," in every hair and in every molecule, because this is how Natural Selection wants things. Organisms are ugly, perhaps, or repulsive, but they are clever utilitarians. This claim weighs heavily in all television documentaries we see about wildlife, in which sometimes fascinating scenes have been overlain with commentaries that ignore beauty altogether and leave room only for animal economy and opportunism. The paradoxical result is that the theory that denies purpose and intention offers us a cloyingly finalistic and anthropomorphic "reality." The end result of such a utilitarian outlook is that intrinsic order and beauty, untrammeled by practical considerations, are completely obscured. All the charm and grace of life, with its torrents of happiness and its unexpected wonders, all the harmony that pervades the created world, have been ruled out by positivistic science because positivistic science uses a different lexicon.

Because of the reservations I have nourished in my confrontations with evolutionism, I have been accused of being a "creationist." I am not; if you will permit, I aspire only to be a creature.

CHAPTER I

ACHILLES INSPIRES REDI

Modern biology can be said to have been founded in the year 1688 (though the term "biology" dates from the early 1800s, when it was first used by Lamarck). In that year, Francesco Redi, doctor, physiologist and poet of Arezzo, demonstrated that worms do not come from meat that has gone bad but from eggs that flies deposit on the meat. Worms that hatch from those eggs penetrate the meat and make it decompose, then they develop into flies. A somewhat sad beginning for the new science, with the stench of death and the swarming of worms and flies. Biology rises from life at a minimum, life at the frontiers of death, insignificant and despicable, then science accords it dignity and autonomy. *Omne vivum ex ovo* ("all life comes from the egg") was the watchword of this new biology as it went about recruiting that ignoble worm and troublesome fly to its purposes.

Redi performed a perfectly simple experiment. He placed pieces of fresh meat in glass jars, some of which he covered with several layers of gauze, and some of which he left open. The meat in the uncovered jars immediately became infested with flies and in a few days was swarming with worms and gave off a putrid stench. The meat in the covered jars remained unaffected for a long time. To be sure, this was not enough to demonstrate that every living thing is born only from its own kind (and Redi himself doubted this was the case with gallflies). Here was a charming fable, a fairy tale about unsullied purity and a recipe for preserving meat. Most importantly, this was a case of delving into the invisibly small in search of the mystery of life.

Why was it so important that the flea should have come out of an egg and not simply from filth? Was it so important that biology should start—almost literally—looking for fleas on nature? The problem was to decide whether life at its minimum was of so little account that it could be "spontaneously" generated and then cross without a passport over the frontier of animal existence. And the answer was No.

After Redi had left the scene, worms were accorded a place in the zoo, with instructions to reproduce by means of eggs, just like birds, reptiles and fishes—and this at a time when it was still undecided whether human females and other mammals also had eggs. In the twentieth century these minuscule worms, which it was conceded could be governed by the same rules as the higher animals, were to rise in importance, as scientists began learning rules of behavior and structure from these minutest of pariahs among living beings. From the mid-twentieth century on, any self-respecting biologist has had to turn his attention to microbes, viruses and molecules in order to be regarded at all and obtain funding, not to mention win a Nobel Prize.

Spontaneous generation has continued to be disputed territory between the "major systems" in biology until more or less our own day. The problem in the eighteenth century was not whether flies came from formless matter, but whether life originated from a seething mass of blind, aimless molecules, or from an intention—from chaos or from a causal principle. For a long time, advocates of chaos sought to create life in a test tube, because if life came from chaos this should be possible. But if life is ordered according to principles, one cannot create it by randomly combining molecules in a test tube.

Biology has advanced in status with every new confutation of the spontaneous generation thesis.

At the same time Redi was going about his task, a modest tradesman in Delft, Anton van Leeuwenhoek, was using optical lenses to scrutinize some of the world's tears—drops of water, blood, sperm. Just like Galileo's lenses, Leeuwenhoek's peeked through a keyhole into the invisible, and kings, emperors and czars soon bowed down and squinted to look through it, too. The secrets of nature began to emerge from the realm of the inscrutable, and life was found residing in nature's minutiae, magnified by the Dutchman's lens. From that time on, naturalists looked at nature with one eye only.

Some time after Redi's experiments, which had not done much to merit a place in history, Leeuwenhoek's optical lens revealed extremely small forms of animal life in agitated motion in some plant infusions, whence the name "infusoria." Protozoans took their place in biology. At first, the infusoria seemed to bear out the idea that life came from matter in chaos, contradicting Redi. Could anyone have found an egg in an infusorium? The idea that living things were generated from unorganized matter—"spontaneous generation"—started gaining adherents again. It became necessary for a far mightier authority to refute the idea and give fresh impetus to the principle that living things must be generated by living things.

It might be fairer to say that modern biology had its beginnings a century after Redi, when the great Lazzaro Spallanzani carried out his well-known experiments designed to disprove the spontaneous generation of infusoria (1765). Spallanzani was born at Scandiano, in Emilia, Italy, and he studied law at Bologna. His passion was the sciences, though, and it was in the sciences that before long he was to achieve exceptional stature. He entered holy orders and taught logic, metaphysics and Greek and, finally, natural sciences at Pavia. His first work was a long letter to Algarotti criticizing the latter's translation of Homer, but he became famous when he published his "Saggio di osservazioni microscopiche concernenti il sistema della generazione de' signori di Needham e Buffon" ("Microscopic observations concerning the system of generation as posited by Needham and Buffon," Modena, 1765).

The Comte de Buffon had a speculative and imaginative mind. Twenty years older than Spallanzani, he had been elected while still very young to the Paris Academy of Sciences, and he had been made Superintendent of the Royal Botanical Gardens as well. A hundred years after Redi's experiments, Buffon boldly maintained that worms, flies, caterpillars, fleas and millipedes all came from putrefaction through spontaneous generation. Unlike Redi, he had conducted no experiment whatsoever; his assuredness derived from philosophical considerations on the possible. He divided species into "noble," "less noble," and "inferior." Among the noble he placed man, whom he considered unique and immutable, and the horse—not quite so noble, what with the donkey not far away. Among inferior beings he placed the insects, which were so variable as to be confused among themselves, and he argued that they were generated from slime and rotting matter. Buffon is looked on as a precursor of evolutionism; and, in fact, he was the precursor of that pernicious approach to science that consists in giving way to the imagination, to the "could be," to the metaphor. Buffon conceived of hippopotamuses and elephants emerging from the primordial slime in the primal epoch when nature was seething with creative potentiality, where they were supposed later to have left their huge fossil bones for us one day to marvel at. The unfortunate insects were the itch of the earth and attendants of death. This boldest of speculators dragged into a nascent biology medieval conceptions and the philosophy of Aristotle, but shorn of their religious component and the rigorous logic of which they had been the beacon.

By contrast, Turberville Needham, an English priest, was a painstaking observer and experimenter. He offered the support of experimental demonstrations to the philosophical doctrine of spontaneous generation, which the times were loath to abandon. Needham put mutton broth in a hermetically sealed jar and heated it on the embers for a good half hour. For a time no sign of life was observable, but before long the broth became a seething mass of minute animal life—infusoria—which could only be there as a result of spontaneous generation. This was no result of mere reasoning, for there was experiment and observation to show it.

Spallanzani repeated the identical experiment except he prolonged the heating, which Needham had reasonably wanted to avoid for fear of compromising the "generative force" of the infusions or altering the air inside the jar. For Needham (and for Buffon, who supported him), Spallanzani had simply destroyed the necessary conditions of life. But how could one know how many hours it might take to destroy those conditions? In reality, experiments are often little more than a liturgy of celebration for a preconceived theory. Spallanzani's theory was *omne vivum ex ovo*; and he concluded from his experiments that the infusoria contained "small eggs" able to resist the limited heat Needham had applied. But there is no such thing as an egg of an infusorium. More correctly, then: *Omne vivum ex vivo* ("all life comes from life").

Spallanzani's experiments on animal fertilization are still famous. Our learned cleric worked with toads and frogs. The batrachian female deposits the eggs in a long mucotic filament—spawn—while the male lies astride her back and clasps her with his front paws under her armpits. What is the function of the male's clinging in this way? Swammerdam, among others, held that this was a way of spreading semen over the spawn in a rapid jet-like motion. The great entomologist Réaumur failed to see anything there, however, despite his assistant's assurance that she had surprised the male expelling something like a puff of pipe smoke from his hind parts. Réaumur had the ingenious idea of dressing the male in taffeta breeches, complete with suspenders; but he saw nothing interesting and concluded that the male did the fertilizing with his claws.

Spallanzani repeated Réaumur's experiment but with greater diligence, so much greater that he was able to discern clear liquid droplets in the male frog's breeches as he was clutching the female. If the drops were laid on the spawn, even with the tip of a needle, it was possible to initiate egg division and development. Spallanzani had no equal when it came to ingenious solutions. He went on to examine the droplets under the microscope, and there he saw myriads of seething spermatozoa. Some of these droplets he diluted greatly, yet they remained effective. He preserved them, dried them and heated them to 35° C, yet they retained their fertilizing power. The droplets lost their power, though, if they were filtered through seven layers of filter paper. Yet when the paper was squeezed, the squash still had fertilizing power. From all this Spallanzani came to the incredible conclusion that the tiny animals in semen had nothing to do with fertilization: The tadpole was there already in the egg, pre-formed, and the spermatozoa had no function whatsoever. Perhaps he was convinced of this because the semen, when diluted to such an extent that nothing could be seen under the microscope, still retained its ability to fertilize; but most of all he believed it because of his faith in *ex ovo*.

Half a century after Spallanzani, Prévost and Dumas repeated his experiments with the same results. They attributed to their illustrious predecessor the experimental discovery of the fertilization of the egg by the sperm, though Spallanzani had actually ruled this out in spite of the evidence he had adduced. In the late nineteenth century, Loeb provided experimental data that would have delighted Spallanzani, showing that sea urchin eggs could develop without the help of sperm (parthenogenesis). Not many years later (in 1910) Bataillon stimulated parthenogenesis in a frog egg by pricking it with a dry needle.

Let us turn back to the confutation of spontaneous generation. Those who held to the theory did not give up the struggle just because of Spallanzani's infusoria. A century after Spallanzani, Louis Pasteur demonstrated that even the minutest bacteria could not originate spontaneously in broth. Pasteur repeated Spallanzani's experiments with practically invisible microbes. Pasteur's adversaries did not sufficiently sterilize the bottles they used, whereas Pasteur performed this task so thoroughly that his work proved to be the foundation of modern sterilization technique, bacterial culturing and the entire science of microbiology. Pasteur was convinced that under normal laboratory conditions life is not generated from non-life; yet he did not fully set aside the question. He thought that if one could provide conditions favorable to the molecular asymmetry of organic matter, the latter could organize itself into life. He put his idea to the test, but he was never able to prove its validity.

Darwin, though a great admirer of Pasteur, regretted that the Frenchman had denied spontaneous generation. "If it could be demonstrated," he was to write to Haeckel in 1873, "this would be very important for us."

Those continuing to hold the theory of spontaneous generation dispensed with demonstrations. Pierre Larousse's 1872 *Grand Dictionnaire Universel du XIXe Siècle* declared that spontaneous generation was a "philosophical necessity," which did not depend on observations and manifestly impossible experiments. The only people to question the theory were the poor physiologists, "blinded by the tradition of dogmatic science." If we apply Karl Popper's rules—which have been adopted into modern epistemology—Larousse was right. The statement "at least in some cases spontaneous generation occurs" cannot be refuted experimentally by rummaging in all the nooks and crannies of the universe. But Larousse was wrong to accuse those who rejected spontaneous generation of being blinded by tradition. The tradition of "dogmatic science," with Aristotle as its High Priest, held that the spontaneous generation of the least of living things was a philosophical necessity.

There will only be one way to refute spontaneous generation. That is to take note of the astronomic complexity of the simplest organisms, and to show that the minimum conceivable life form calls for structures so elaborate that no fortuitous accident can bring them all together. But we had to wait until the second half of the twentieth century for the proof.

What was it then that inspired Francesco Redi to perform his now famous experiments on the generation of worms, and thus place himself at the head of the retrograde, the "blinded by tradition"? It was certainly not Aristotle, nor medieval tradition, nor popular belief—all wedded to the idea of spontaneous generation. Nor, again, was it the Bible, in which the Samson story describes bees coming out of a lion's carcass.

Curiously, and surprisingly, Redi acknowledges in his *Experimenta circa generationem insectorum* (1668) that the idea of looking for the causes of corruption in meat came to him from Homer's *Iliad*. After reading it he began to wonder whether "all those worms in the meat came from the seed of the flies alone and not from the putrefying meat itself." The passage that started Redi thinking is in the opening lines of Book XIX, where a mourning Achilles is speaking to his mother, Thetis, beside the body of Patroclus, who had been slain by Hector. Achilles says:

> "… a fear oppresses me
> lest, now in Patroclus' wound,
> some vile insect should enter,
> a generator of worms, and lest the corpse (alas, lifeless)
> should rot, yea and all become putrid."

Thetis then assures Achilles that she will keep the swarm of flies away, so that the body of the hero can remain uncorrupted for a year and become even more beautiful. So Redi kept the flies off by means of gauze coverings, and the meat remained uncorrupted for a long time.

The true ancients evidently believed in the self-reproduction of lower forms of life, a concept excluded from Greek philosophy and medieval thought for two millennia. Every foundation, in all cosmogonic myths, begins with the egg that opposes chaos. The egg contains a principle of organization from which all differentiations arise. The cause of birth and rebirth, the egg holds the germ of future reality. At the side of Patroclus' corpse, how could Achilles have thought that the handsome hero lying there before him would generate worms and swarms of flies? For the thoughtful hero, each being has its own species and its own life. Achilles, except for that unfortunate heel of his, was of divine species.

CHAPTER II

FROM AN EAGLE'S EGG, AN EAGLE

Y sobre el desorden del mar...
volò volò volò volò
su equatiòn pura en el espacio.

(And over the disorder of the sea…
it flew flew flew flew
its equation pure in space.)
–From "Cormoran" by Pablo Neruda

The greatest problem in biology for centuries—for millennia even—has been: How does an egg turn into an embryo? Take the human egg, the size of a speck of dust and known to science only since the early 1800s. As this egg develops, the fundamental plan of the embryo becomes established in a short time—from one to four weeks following fertilization. During this phase a homogeneous lump of cells turns into an animal in miniature; the tiny animal then becomes recognizably human. But the real mystery of morphogenesis, and of this life of ours, is compressed into those three weeks when the spatial organization (or "regional specification") of the organism emerges.

Most eggs are globe-shaped. The frog's egg, the one most studied in this regard, is no larger than a pinhead. Under the microscope many eggs exhibit a polar region, with the nucleus and various planes of substance stratified in relation to the pole; and there these lie, one would think, with no intention of forming an animal. As the egg matures, the chromosomes become discernible in the nucleus. After a pair of cell divisions, these are reduced by half (the rest being ejected from the egg), and the remaining ones are ready to welcome the other half from the sperm that is about to arrive. Following fertilization, the chromosomes will show themselves at every division, in pairs.

When chromosomes were discovered in 1875, biologists soon realized that they were the key to solving many problems—except the problem of development. Chromosomes split in regular and symmetrical fashion at each cell division and eventually distribute themselves uniformly in all cells, where they maintain the same number and the same forms. Each of the cells of an organism—whether in the various embryo layers, in the liver, hand or heart—has exactly the same chromosome set. This uniformity is quite incapable of explaining the distribution of functions and forms in the embryo. True, it has been shown that in different cells different chromosome regions are in operation. In some cells chromosome activity manifests itself in localized puffs. Nevertheless, these manifestations are not due to spontaneous initiatives by the chromosomes themselves, but are responses to stimuli originating in the region of the embryo where the nuclei (with their chromosomes) have ended up.

In an attempt to understand regional differentiation in the embryo, research has concentrated on two phenomena: the spatial distribution of the egg cytoplasm; and "induction," which is the effect of local stimuli (external or internal) on the developing embryo. The geography of cytoplasm has been studied chiefly in the eggs of invertebrates such as the sea urchin, while induction has been studied in the embryos of vertebrates such as the frog. In this latter case experimenters have been able to mimic certain stimuli with a needle, micropipette, thread, or tip of a lancet, and they have ended up with instructive monstrosities in the process.

The problem of morphological development is first of all a problem of asymmetry. Physicists have a law (P. Curie's principle) which states that, if a phenomenon exhibits a certain asymmetry, the same asymmetry will be found in at least one of the conditions determining the phenomenon. If a body has a front and a back, the cause, too, must have a front and a back. But chromosomes have no such asymmetry; they have no back and no belly, no right and no left.

If you want a baby to understand right and left, there is no point in shouting at him. You can explain to him, and have him adopt the distinction, only because right and left are already present in his instinct. And then you will tell him that his right hand is the one he uses to hold his spoon, or, when he is a little older, the hand with which he makes the sign of the Cross. The "right"/"left" message does not teach anything unless it fits in with some pre-existing bilateral asymmetry in which two opposing and different half-worlds can originate and develop.

Embryologists of the late nineteenth century thought at first that they had discerned elements of asymmetry in the egg cytoplasm, this being readily observable in many invertebrate species. The ascidian (*Styela*) egg is as transparent as glass and reveals within its interior pigmented granules arranged in strips, concentric strata and crescents. The pigmented granules run about and shift aside when the egg begins to develop, and they take up their positions in specific regions of the embryo. The organism thus seemed at first to be predestinated in the egg's geography.

Yet research along these lines came to nothing. In ascidian eggs, just as in the eggs of other species, a brisk centrifugation could bring about shifts in the pigmented areas. A yellow half-moon band was displaced and ended up in different cells following segmentation; yet the embryo developed quite normally. Some thought that the intrinsic differentiation of the cytoplasm was something stable and built-in, and that the pigments were merely secondary manifestations of it. But by then the pigment game was lost.

The spatial frame of the future organism was then entrusted by some embryologists to the egg cell wall, or cortex, which does not become deformed under centrifugation. To distinguish this from the endoplasm, or the cell's central cytoplasm, the part adhering to the cortex was given the name of ectoplasm—a sort of ghost leaning against the wall.

An experiment conducted by Wilhelm Roux in the late nineteenth century seemed to confirm spatial distribution within the egg of the future areas of the organism—a mosaic of local destinies. Roux worked on frog's eggs at the stage of the first two cells (blastomeres). With a needle he killed one of the two cells, and the other cell developed into a half-organism with one side only. This was interpreted as proof that form was a matter of distribution of cell material.

Hans Driesch (1892) introduced a variant into Roux' experiment by separating completely the first two blastomeres of a sea urchin egg. From each of them—i.e. from each half-egg—a larva (pluteus) developed that was perfectly normal, only smaller. The same result was obtained if the egg was subdivided after several cells had formed. From each cell a "twin" developed that was identical to the others and (except for size) to the larva that would have formed from an egg that had been left alone. So there was no preformed mosaic.

It sometimes happens that from the latest failure or disappointment a lost thread re-emerges—the elusive figure we had sought in vain. The fact that an egg can be divided into two half-eggs that develop into the same form after separation can be expressed by a magnetic field analogy. In the egg there is a "field" with this strange property: Cutting the egg in half produces two fields that are identical with the field of the original egg and with each other. Cutting the pieces again reproduces the whole field two-, four-, or even eight-fold. Something similar happens with a magnetized iron bar, with its magnetic field lines revealed by iron filings. Break the bar in two and place the halves at a distance from each other, and the iron filings will show that each piece forms around itself a field similar to the original one. Just as the original bar had a south pole and a north pole, so also the broken pieces each have a south pole and a north pole.

Driesch called the egg's field the "morphogenetic field" to indicate that it has the capacity to generate a form; but it does not have, nor is it, a form. It is a body-less structure, an immaterial energy flux.

Pavel Florenskij, a Russian physicist and theologian (1882-1937), imagined a similar field on the surfaces of icons that portray sacred images. "The essence of the surface is dormant until it is evoked; but once paints are applied to it these awaken it ... just as a magnet's invisible lines of force become visible thanks to the iron filings."

Driesch's "field" has not only the property of forming the organism but also that of restoring it to its proper form following any perturbation (self-regulation). Every single part possesses morphogenetic properties higher than that which it will express (equipotentiality). We have seen how, for a short while, embryo cells remain totipotential. As development proceeds, totipotentiality tends to decline. The field is a morphological project that gradually differentiates and becomes regionalized. Such a procedure does little to explain the precision of the end result, of the perfected form, the marvel of nature. Driesch rounded off his field's properties by assigning a causal role—autonomy, almost—to morphological outcomes. An identical final structure can be arrived at via different avenues of development, as though it beckoned morphogenesis to a stereotypical end-product (equifinality).

At the stage when the frog embryo starts to change shape and involute (the gastrulation stage), the destiny of the embryo parts is not yet decided. At that point, if a small lip of tissue that is readying itself to become a neural plate is transplanted into one side of the embryo, the transplanted piece will alter its own development to become part of the epidermis on the side where it is now lodged. If the lip is transplanted only a little later, the result will be quite different. At the site of the transplant a neural plate will form, and this will bury itself under the surface, giving rise to a neural groove and then to a neural tube, which can go on to form eyes and ears and a spinal column with its associated muscles. Finally, a head, with the beginnings of a back, will emerge from the side of the embryo which has now become monstrously bicephalous. In this case the cells of the neural lip are said to have "induced" a secondary, ectopic neural plate. The transplanted lip, it is also said, contains the frog's "organizer," since it triggers the processes that form the neural tube and eventually the tadpole.

So, does this small piece of tissue (the "blastopore lip") contain the secret of differentiation? Does it tell the embryo what to do? The answer is: No, it does not, for even when killed by heat it induces a neural tube. In fact, one can replace it with the "lip" of another species or with a saline solution, and organization proceeds anyway. The inducing agent bears no information regarding form. It is simply a non-specific evocator, possessing none of that magic and complexity which had been expected of it. Lev Beloussov writes: "The events, highly specific and regularly localized, of biological development can be generated … by dynamic, entirely degenerative, and vaguely dislocated perturbations." The organ's asymmetry lies not in the immediate causes or circumstances. It is an innate property, indeterminate and everlasting, of a "self-organizing" organism.

The term "field" is perhaps too rustic, poetic and enchanted to express what the morphologist means when he uses it. From the very moment it receives its first signal inside the egg, the field begins to change, to move, to become more complex. It is a "system undergoing transformation," a whirlwind; and what appears to the observer is but a momentary pause in its performance—sometimes a pause, sometimes the end. The field is like the Spirit of the World that, in Goethe's *Faust*, catches the doctor up in the vortex of its breath and twists him frighteningly like a worm.

> "In the stream of life,
> in the tempest of facts
> I go about, wandering
> upward and backward!

Birth and burial,
an eternal sea
an iridescent fabric
an incandescent life,
thus I bend to the rumbling
loom of time
and fashion for the godhead
a living garment."

An open flower, the flower of the painters, is produced by the transformation of a seed and a bud on the way to becoming fruit. It is also the transformation of a leaf whorl, as Goethe saw things. It may be the transformation of a tiny wayside flower in the hands of a plant breeder—or of some ancestral flower, as Darwin imagined. The flower is also a solution in development of a mathematical formula, as D'Arcy Thompson used to describe it; and finally, the unfolding of a "morphogenetic field," or—why not?—the metamorphosis of a princess in a tale of magic.

The flower in the fairy tale turns back into a princess at the touch of a kiss; but how does the botanist's flower, offspring of so many transformations and upheavals, produce in its pistil the tidy oosphere where the journey begins that will lead to other flowers? How does an animal egg, once its field has been turned around and scrambled, reproduce the intact and olympian miniature whence everything has its beginning?

The enigma was solved (with a touch of histological approximation) by stating that the egg that gave rise to a newborn child was not produced by a bride in love, but budded off from the cell series that produced the lady herself in her grandmother's womb. Egg from egg. It did not have to retrieve the spotless beauty and totipotentiality that the parent egg had dispersed in the developing organism, for the simple reason that it had never lost it.

German cytologist August Weismann proposed that egg derives from egg via the "germ line," specialized cells that remain separate from the other cells of the developing organism. The organism is a sort of lateral expansion that in no way shares in the uncontaminated germ line joining egg to egg (or, in the male, egg to sperm). The germ line establishes a moat, a barrier that shields genes destined for the next generation from experiencing the world. Weismann was criticized for having segregated heredity away from life, and for having dissociated it from morphogenesis. In my view, beyond that same barrier even the "morphogenetic field" of the egg is shielded from the ups and downs of life's relationships. It holds itself static, totipotent, virginal, allowing itself to be caught up in the tempests of becoming, but all the time keeping a portion of its undiminished essence within the walls of a secluded garden.

According to a theory of Darwin's that is little known today but was dear to his heart—the theory of "pangenesis"—an egg is made from features of the parent organism that transmit their earthly past through the seminal fluid in the form of little particles. According to pangenesis, the entire organism generates the offspring. Only in this way could Darwin explain the evolution of the species—i.e., as a decanting of the vicissitudes of the parents' lives into the offspring. For Darwin, evolution was the cumulative experience of the world's organisms over time. He got this idea from his illustrious, unappreciated French precursor, the Chevalier de Lamarck. Before Darwin came along, Lamarck proposed the theory of the transmission of acquired characteristics. The transfer of worldly acquisitions from the environment to offspring was a sort of spontaneous generation of life from non-life, and this was evolution. Darwin never thought that evolution was anything else, and he would have disavowed the Theory of Evolution propounded in his name in the twentieth century.

Once the effect of the environment is excluded, whence, one may ask, come the differences between living beings? Weismann suggested that the differences must have been present in the first beings to populate the earth. Species differentiate among themselves because of something received from distant ages, remaining intact for millions of years, unreachable by influences of the body and apart from transactions with the environment. To Darwin's "pangenes," coming from all parts of the organism to form the germ of each generation, Weismann opposed his "biophores," present from the beginning of life and preceding all organic forms, including eggs. He maintained that these "determinants," as incorruptible as ideas, were transmitted via immortal germ lines to produce bodies, again and again, as glorious and mortal by-products.

Life thus returns to its origins or, rather, holds on to its origins by clinging to the handrail of the germ line.

Samuel Butler expressed Weismann's theory in the following terms: "The hen is the means whereby an egg constructs another egg." This evokes a barnyard scene where the hen is a gossipy creature, incapable of flight and good only for laying eggs. The hen well expresses the uselessness of the organism, apart from her function as a bearer of eggs. But Weismann also said: "From the eagle's egg, the eagle."

The shells of an eagle's egg and a hen's egg are barely distinguishable. Their egg cells, nuclei and DNA look almost identical. And yet from the hen's egg there hatches the chicken, and from the eagle's egg the king of birds: Powerful and immense, with hooked claws, imperial head and great square tail, it soars aloft, its outstretched wings at times motionless, at times stirring in solemn strokes, the feathers at its wing tips separated and curving upwards.

From an eagle's egg, an eagle.

CHAPTER III

THE ECLIPSE OF THE ORGANISM

The final quarter of the nineteenth century witnessed a decisive turning point in the science of living things, when chromosomes ("colored bodies") came on the scene and claimed center stage in biology. Chromosomes are tiny rod-like assemblages measuring a few microns in length, which become visible within the cell nucleus at the time of its division. They are found in plants and animals, and they occur in the same number and the same form (with minor exceptions) in all of an organism's cells and in all organisms of the same species. Their diagrammatic representation (idiogram) was from the start taken as a specific identifying feature of the species, rather like the bar code a cashier scans at the supermarket. Within the cells of the body, or soma, they occur in pairs. Fruit flies (*Drosophila*) have 4 pairs, humans 23 pairs, and soft wheat 21 pairs. Closely related species have chromosomes that are similar in both number and form. Thus, humans have 23 and chimpanzees have 22; donkeys have 31 and horses have 32; zebras have 16—making a horse the sum of two zebras. Soft wheat has 21 pairs, hard wheat has 14 pairs, and some wild wheat species have seven pairs.

The regularity of certain chromosome series, particularly in plants, seemed to suggest a relationship between chromosomes and form; but the series in question were limited to closely related groups with similar features, and thus provided no key to explain increasing complexity.

Between chromosome number and the evolution of species it was immediately clear that no clear relationship exists. The number generally ranges between 16 and 25 pairs per cell nucleus; but a roundworm (*Ascaris megalocephala*) has only one pair, while a fern (*Ophioglossum petiolatum*) has 150 pairs.

Following the discovery of DNA (of which the genes in the chromosome are made), an attempt was made to establish a relationship between the amount of DNA in a cell and the "evolutionary" complexity of the organism. What was found, indeed, were two levels of DNA: the DNA of bacteria, with their millions of pairs of nucleotides, and the DNA of higher organisms, with their billions of pairs. The difference was not directly related to the amount of genetic information contained in the cell but to the organization of the chromosomes. Bacterial DNA is a string of genes; the DNA of higher organisms includes long non-coding sequences between and within genes. Since these sequences were non-coding they were dubbed "junk" DNA. The attempt to establish a relationship between amounts of DNA among different groups of animals proved disappointing. Mammals have around five billion pairs of nucleotides, reptiles around three, birds around two, and fishes between 0.3 and 3.0 billion pairs. So far so good. But what about amphibians, which have about ten billion pairs and in some cases 100 billion—a level one finds in several dipnoid fishes as well? The mollusks have levels of DNA like those of the vertebrates, and worms like those of the birds. Flowering plants waver between two and 500 billion pairs.

DNA, as we said, is the stuff genes are made of, and, since genes are directly involved in metabolism, it was thought that a gene count might offer a better index of organismal complexity. Anywhere from three to five thousand genes have been counted in bacteria, 6,000 in yeast and 25,500 in one of the *Cruciferae*, 13,600 in the midge and 26,000 in *Drosophila*. Man, in whose genome should have been found his entire civilization and his destiny—the Parthenon and the Ninth Symphony—has

25,000–30,000 while *Cenorhabditis elegans*, a small worm 1 mm long with only a thousand cells has almost 20,000. And where did all this lead? To the conclusion that biochemical complexity has little to offer in explaining evolution.

The story of the evolution of organisms, told in terms of chromosome numbers, or numbers of genes, or amount of DNA, was a complete failure. The stance taken by biologists, therefore, was to discount differences and to concentrate on "universal" DNA. In the second half of the twentieth century, DNA—its structure, its self-replication, its codes, its exchanges and its decay and repair—became the central interest of biology, while organisms disappeared below the horizon. In many papers on DNA the organism is barely mentioned, for it no longer showed forth the glory of the Lord—or showed only His propensity to gamble—leading Salvador Dalì to exclaim: "And now, this announcement of Watson and Crick's is the real proof of the existence of God"—a universal deity that presided over essential life but was uninterested in vain morphological variations.

This was how, halfway through the 1900s, molecular biology was born, pledged to study the DNA of genes and its primary products, the proteins. Beyond the proteins, between proteins and final forms—the wilderness.

Molecular biology took on such dimensions and gathered such momentum as to outclass the other biological disciplines, and even chemistry and physics.

The astonishing thing about molecular biology was not so much the knowledge that was emerging from the study of macromolecules but the fact that living nature showed itself to be so accessible, so ready to yield up its secrets without any modesty or reticence. It seemed as though life could be disassembled and reassembled like a child's blocks. Some people then placed their faith in the omnipotence of biology and in the prospect—it seemed only a matter of time—of being able to put life together and change it in a test tube. To be fair, nobody today expects to be able to construct a cell or a homunculus in vitro, and genetic engineering (as it has been called) does not erect biological edifices, but limits itself to simply adding a touch here and there.

One of the fundamental principles of molecular biology (now enshrined as the "Central Dogma") assigned to DNA the role of absolute governor of the life and inheritance for the cell, and consequently for the organism. The Central Dogma proclaimed: DNA reproduces itself and produces proteins; proteins do not reproduce themselves and are unable to modify the DNA that encoded them. In other words, the information proceeds from DNA to DNA and from DNA to proteins, but it never makes the return journey from proteins to DNA. Such a theory excludes peremptorily the possibility that the vicissitudes of proteins could have any effects on DNA—i.e., that acquired characters of a Lamarckian sort might enter into the hereditary circuit.

The Dogma seemed to offer a molecular version of Weismann's theory of the germ line, which holds that the germ is the origin of a new germ and of the soma, but the soma does not talk back to the germ. The egg makes the hen; the hen doesn't really make the egg—she merely lays eggs that derive directly from the egg that made her. In the new molecular version, DNA was the egg and proteins the hen.

As a corollary to the Central Dogma, we were given a revised version of the Theory of Evolution. The DNA "text" can undergo chance mutations ("copying errors"), and these alter the texts of proteins with the result that they become better (or less well) adapted. The best adapted—the fittest—survive and dominate, while the others disappear from the scene along with the genes that produced them. And this is Natural Selection—which is to say, molecular Darwinism.

In Darwin's theory of pangenesis, offspring come from all parts of their parents. The foal is a horse because it is the scion of a mare and a stallion. For Weismann, however, the foal is a horse because it is the issue of a germ line that produced the mare and the stallion in the first place—a younger brother, shall we say. Descendants take after their parents because they derive from the same germ line. In his theory of heredity Weismann is an anti-Darwinist; moreover, so is molecular biology.

Unlike the latter, however, which shares with Darwin an aversion for the world of ideas, Weismann was an idealist. His germ line defends the germ against outside happenings, but Weismann's germ is not made up solely of particles and scraps. No, the germ contains immaterial "ideas" that control the development of the organism in the direction of "type" forms. Weismann was imbued with nineteenth-century German idealism. His determinants are inaccessible, like the particles in the germ cells and ideas in the Platonic empyrean.

In the official portrait of August Weismann (as Franco Scudo pointed out to me), the austere professor of Freiburg im Breisgau, all but blind from the age of thirty, wished that two butterflies be given a prominent position. These were two forms of *Vanessa*, one with yellow, the other with reddish wings. They came from the same hereditary line, expressions of the same idea, the same species—the one, *prorsa*, deriving from summer larvae, the other, *levana*, from autumn larvae that had passed their diapause. The fairy-like pair provided an illustration of the concept that from the same germ could emerge different forms, and therefore that the germ could not be expressed simply as a collection of predeterminants. The organism and its habitat both contribute to its expression.

Molecular biologists would be much less fascinated by the vagaries of butterfly wings and by other traits of the soma. Once it was proclaimed that "everything is written in the nucleic acid message" (Jacob, 1974), the rest began to decline in importance and interest. Biology took to deciphering from DNA general instructions for use, and forgot about "uses"—that is, the real world of living things.

In the early years of my scientific work, one could perceive among biologists of abstract life a genuine hatred for form. For those biologists, form was outdated and redundant. I heard them deride botanists who describe the profiles of leaves and zoologists who take an interest in lions' manes. Since biology no longer wished to be considered a natural science, but strove to be an exact science, its task was to deal with the vital minimum that brought scientists close to the origin of life.

Molecular biologists were no longer interested in forms, differentiations and organizational plans, since they were convinced that their material would in the end provide an explanation for everything. We can say that they began to lose interest in life and organisms; and those who still clung to the singularity of nature and to forms were classified as "vitalists" or "organicists." Not only did the new biologists distance themselves from interest in the totality, but they also applied a reductionist approach to method. As Jean Piaget put it so clearly before DNA came on the scene in 1946, a problem is scientific when people have succeeded in isolating it in such a way that its solution does not call into question the whole; otherwise it remains merely a philosophical problem. The eclipse of the organism coincided with what elsewhere was seen as the eclipse of philosophy.

What really happened with the advent of DNA (and this is the rule with all scientific revolutions) was not that a solution was found for the existing problem of biological form. Rather, that problem was abandoned for another one: the problem of the material transmission of heredity. And so it was that interest turned to asking how "determinants" of characters passed from generation to generation. But characters, and even form, ceased to be of interest. These became "markers" of hereditary factors through whose agency it became possible to follow chromosome rearrangements, gene recombinations, and the fate of DNA. Little by little the experimental organism became smaller and smaller; it lost body and character, passing from Mendel's garden peas to Morgan's fruit flies, then on to molds, yeasts, bacteria and viruses, and finally to DNA in vitro.

I myself have manipulated bacterial DNA. This is extracted from a cell culture. The cells, with their walls removed, are precipitated in alcohol and gathered up again, rather like a thin thread, and wound round a glass rod, and then dissolved once more. The DNA so extracted is then mixed with a special DNA vector (plasmid); it is treated with a "restriction" enzyme which cuts the vector and DNA into pieces; and these pieces are then recombined by means of another enzyme (ligase).

The DNA thus conveyed on its vector is then added to a cell culture of a strain that is different in character from the original strain. The treated cells are plated onto a solid, selective medium on which the recipient strain is unable to grow, so only the few cells that have incorporated the appropriate "foreign" bit of DNA bound to the vector go on to form colonies. This is the procedure followed today for laboratory exercises, and from it has developed the technology that we now know as "genetic engineering."

First the attempt was made to produce human substances such as insulin and growth hormone from engineered bacteria; later to put healthy genes into patients who lacked them (gene therapy); and then to transfer foreign or human genes to laboratory animals or to farm livestock (transgenic animals). Plants were particularly suitable to the enterprise and lines of genetically modified organisms (GMO) are now invading the world.

Man was involved in a series of Faustian projects. The molecular biologist thought he could surmount barriers imposed by nature and, as the saying went, take over control of evolution. What did it matter that evolution studies were in the doldrums? Man had taken hold of the destiny of nature and of the world by showing that evolution was within his reach. Now we had bizarre theories of evolution, in which novel techniques were offered to nature so it could use them for its own improvement. Just as Darwin thought in terms of the magic wand of the animal husbandman, so the molecular biologists imagined that miracles would come from that magic glass probe with DNA wound around it.

We can conclude by noting that none of the manipulations making up genetic engineering has succeeded in taking off, and that nature has effectively defended its frontiers. Hundreds of biotechnology firms have set up shop, have been quoted on the stock exchanges and have taken out industrial patents. Biology is awash with billions of dollars such as it had never seen before. Science has taken on the great wager… and lost. For decades it has sold plans, illusions and deceptions. The achievements have been negligible, and the prospects are scarcely encouraging.

Nor should this surprise us. The rule in science is that the practical conquests of technology are rarely the fruit of theoretical knowledge, but rather that practice precedes and determines theory. DNA could never construct the objects from which it had disdainfully distanced itself. When abstract science gains the upper hand, technology goes into eclipse. The great innovations that have changed our lives have a very modest scientific content, and the great theories that have changed science were developed on their own account, in a world cut off from reality. "Great technology," wrote Richard Reeves in 1983, "is constructed on the edifice of past technology, with only a little help from science." And Derek De Solla Price comments: "Inventions do not hang like fruit on a scientific tree."

It is said that in 1833 there was great excitement in London. Everyone was flocking to see an experimental five-horsepower "caloric engine" being demonstrated by the Swedish engineer John Ericson. Michael Faraday, then considered the greatest scientist of all time, let it be known that he would give a lecture at the Royal Institution explaining how the machine functioned. On taking the podium, however, after a moment's hesitation he had the candor to admit that he was mistaken: His theory could not explain how the machine worked. The five-horsepower engine proceeded to work anyway, even though the watching crowd was ignorant of how it did so.

I do not think there is even one molecular biologist—I'm sorry to say—who will admit that his theories cannot explain the gait, the functioning, of a horse.

CHAPTER IV

WOBBLING STABILITY

⎯⎯⎯⎯⎯⎯⎯⎯⎯•⎯⎯⎯⎯⎯⎯⎯⎯⎯

Twentieth-century biology was inaugurated by an Augustinian priest in Moravia, Gregor Mendel, who worked for somewhat less than ten years crossing garden peas in a monastic garden. He concluded his experiments in 1865, when he read his findings to a scarcely attentive and rather bored audience at the Naturalists' Society in Brünn. It is common knowledge that his work was "rediscovered" in 1900 by three central European botanists. Mendel concluded from his experiments that determinants of alternative characters (e.g. white or red blossoms) come together when a plant is crossed but remain distinct, and express themselves again among the descendants of hybrids. If a plant with the red flower trait, indicated by the symbol A, is crossed with a plant having white flowers, indicated by the symbol a, a hybrid Aa is formed. This hybrid has red flowers, which means that A is dominant with respect to a. A dominates a—conceals it—but does not obliterate it, so that plants with white flowers reappear in the hybrid's later generations. Mendel described his experiments using algebraic symbols and probability calculations, and these enabled him to tell in advance the frequency at which the different pairs of characters might recombine, and to verify it in carefully recorded experiments. He also calculated that character

(and determinant) frequency remains constant in subsequent generations. In other words, the genetic make-up of populations is invariable,
except where certain possible disturbing factors intervene; but these he
diligently excluded from his experiments.

Mendel's thinking did not appeal to some people in the years when
the Theory of Evolution was taking hold, a theory whose fundamental
postulate was that the genetic composition of populations varied from
generation to generation, and that the accumulation of these variations
over time led to the origin of species—to evolution itself. The affirma-
tion of Mendel's laws in the early 1900s corresponded to the eclipse of
evolutionism. The eclipse lasted until the period between the two wars,
when a sort of compromise was reached between the two theories that
became known as the "New Synthesis" (or Synthetic Theory), from the
title of a book by Julian Huxley.

What really happened was that Mendelism ruled out almost all the
forces that Darwin had invoked to explain evolution. For Darwin, heredity was the result of a mixing of seminal fluids (though critics objected that mixing would have the effect of dissolving the variations). Mendelian heredity consists in a recombination of traits that associate with
each other but do not blend, with the result that a variant trait always
has a chance of reemerging. Darwin's mixture theory was for him no
idle idea; it was a necessary corollary to his conviction that the environment acted directly on the germinal fluids inside the body when these
were preparing to pour into the semen at the moment of orgasm. Mendel's hereditary determinants were not amenable to such influences and
transmissions. They were static, permanent, and fully indifferent to the
environment. In some quarters their stability was considered reaction-
ary—clerical even, given that they came from the garden of an abbot.

Mendel remained unnoticed for 35 years, as long as Darwinism was
in the ascendant, and his laws came to light only when evolutionism was
going through a crisis in the early 1900s. Evolutionism got to its feet
again by clutching on to Mendelism, when little was left of Darwinism

beyond Natural Selection, which in his later years Darwin himself had somewhat revised, and of which genetics took little account.

The accommodation was rather equivocal, because the variations of Mendel had little to do with the variations needed by the Theory of Evolution. Actually, according to Mendel characters reassort but they do not change, either in type or in frequency.

Although Mendel is credited with the discovery of the "particulate" nature of hereditary determinants, he advanced no hypothesis as to their structure. He indicated characters of pure strains by means of the symbol A or a, and of hybrids by Aa. If he had thought of particles or the like as being associated in a varying pattern, he would have used the symbols AA and aa for pure traits—a notation that entered later into modern genetics. The "gene" as a physical entity did not come into evidence until some years following the rediscovery of Mendel. This gave a physical connotation to the Mendelian "essence" but severed its connection with characters. Genetics thereby began to lose its status as a formalized science and gradually became the material analysis of the gene, its molecular structure and the alterations it undergoes.

The variations about which modern evolutionists talk derive from a blind process, namely "mutation," unknown to either Darwin or Mendel. According to the Synthetic Theory, the accumulation of diverse mutations is what produces differences in fitness between species. This, of course, requires long periods of time, plus the operation of Natural Selection, the latter choosing from among the mutations the very rare beneficial ones, the most vital. Mendel had carefully ensured that the strains he had chosen for his experiments all possessed the same degree of viability; otherwise, their differences would have distorted his calculations. In no way could he have imagined that precisely those external disturbing factors should give birth to the theory that was destined to change the world. From Mendel the Synthetic Theory had taken over chiefly the idea of recombination—chance reassortment among characters. This process, a prerogative of sexually reproducing species, seemed as it were invented for the purpose of multiplying variability. Ten pairs of combining characters can produce 2^{10} combinations, or more than two thousand different types.

Give me variability and an efficient method of selection, and I will raise up the world—this long remained the conviction of the evolutionists of the 1900s. And since sexual reproduction seemed to be the multiplier of variability, it was dragged in as an accomplice in that universal transgression that biological evolution proved to be. I recall that I, too, ages ago, had an idea that if I were to grow several billions of bacteria and I possessed a technique for selecting whatever I chose, I could eventually extract an elephant from my Petri dishes.

Recently, new sources have been found to contribute to variability in the hereditary material. "Jumping genes" are able to move from one chromosome to another. Split genes are able to reassemble their own messages. "Directed" mutation has been observed by John Cairns and colleagues in organisms that change their DNA in an adaptive, directed manner.

It sometimes happens in science that, in order to explain some phenomenon, people construct models that have the effect of explaining things away. In the end we find that the phenomenon for which we now have such a good explanation isn't there at all. As observed in nature and reported in paleontological finds, living species seem to be substantially stable over time, capable of resisting change for millions (and in some cases hundreds of millions) of years. Things fluctuate, but only in order to remain what they are.

In Tomasi di Lampedusa's *The Leopard*, a novel about the author's native Sicily, the protagonist is a prince, restless yet reluctant to see the old order changing. His favorite nephew, the devil-may-care Tancredi, seems to be ready to come to terms with the revolutionaries and with Garibaldi's Thousand. Tancredi, "in an excess of seriousness," declares to his disconcerted uncle: "If we are not in on this, they'll decide on the republic. If we want everything to remain as it is, everything will have to change, if you understand what I mean." Here we have, not a dishonest opinion, but a theory of stability, which we may formulate as follows:

Only oscillations deriving from variability enable a species to remain faithful to its type.

A man on a bicycle keeps his balance by constantly wobbling from one side to another, regaining the upright position by means of an instinctive thrust in the direction opposite to that in which he would otherwise fall. Woe betide him if he were to stiffen up or if his wheels got stuck in a trolley track! The cyclist must impart oscillations to the handlebars. Have you ever seen a motionless tightrope walker? All of which shows that stabilizing variability and "wobbling stability" are not contradictions in terms, not oxymorons.

All organisms possess immediate defenses against variation, which go by the name of repair mechanisms. If a DNA helix suffers some damage, a mechanism is ready to remove the damaged part and reconstitute it by copying from the companion helix that has remained unscathed. And how, we may ask, does the cell know which of the twin helixes is the one to copy? No problem: It recognizes the "old" helix as compared with the "new" one that contains the copying errors.

The most effective method whereby organisms defend themselves against the tendency of future generations to decay genetically is sexual recombination. In the 1930s, when the Synthetic Theory took center stage, sexual recombination was thought to increase variability and offer the maximum number of genetic combinations from which selection could make its choice. Yet it gradually became clear that the role of sexual recombination was, if anything, a conservative one. Every group or individual involved in sexual reproduction contributes its genes to the pool of the population, where the waters are constantly stirred as a result of the tendency of individuals to get mixed up in their embraces. The only way that a group can distinguish itself from the crowd is to withdraw from the world like the anchorite and the cenobite, or to depart as a colonizer who sets out for new lands (or new pools) to form an isolated community (allopatric speciation). Another distinction can be made within the overall community by a tiny group that goes in for an eccentric lifestyle, or settles in an exclusive niche, which in either case interrupts the flow of amorous exchanges (sympatric speciation). Sexuality creates confusion and uniformity; it mixes and stabilizes—even if heredity is not made up of Darwinian liquids but of Mendelian grains.

The limited variability that a species maintains, with its partial internal distinctions, its semi-isolations and its general promiscuity, is its guarantee of stability over the long term. Rules and transgressions of rules, promises and infidelities are enough to keep alive the variability that distinguishes and the diversity that rejoices. An excess of uniformity suffocates a species; an excess of diversity disintegrates it.

Sexuality has brought joy to the world, to the world of the wild beasts and to the world of the flowers, but it has brought an end to evolution. In the lineages of living beings, whenever absent-minded Venus has taken the upper hand, forms have forgotten to make progress. It is only the husbandman that has improved strains, and he has done so by bullying, enslaving, and segregating. All these methods, of course, have made for sad, alienated animals, but they have not resulted in new species. Left to themselves, domesticated breeds would either die out or revert to the wild state—scarcely a commendable model for nature's progress.

Modern genetics tells us that it is selection that intervenes to change species, and that variability by itself is of no effect. It is not that vague oscillating-in-order-to-remain-the-same that changes things, as Lampedusa's Tancredi sagely put it. Only Natural Selection, with a capital N and a capital S, the official textbooks tell us, can change forms and species. The example usually trotted out is the peppered moth. Thereon hangs a tale, and it goes like this:

Once upon a time a population of gray peppered moths used to alight on the silvery trunks of birch trees, thus eluding predatory birds. Then along came industry with its smoking chimneys, and the birch trunks blackened with soot so that the moths now stood out—gray on a black background. But, presto! Along came a strain of black moths that were able to camouflage themselves against the dark background, while their gray cousins remained conspicuous and became easy prey for the birds. That (so the story goes) is how gray moths disappeared and the black ones came to prevail. The result was "industrial melanism." The tale even had an ecological moral, and it was adopted as a model of evolution under our eyes: a tale told to children, and given pride of place in our official textbooks.

The fairy tale of the peppered moth is plausible, but untrue. What of the many examples of melanism in regions where there are no industries and no soot? What of the fact that moths are not in the habit of alighting on tree trunks at all but prefer to hide among the leaves? And what about places where human efforts ensured that the trunks regained their original coloring, whereas the moths in many cases have remained black? There is the further point that the moths used by Kettlewell in his experiments were caught in lighted traps that blinded them and stunned them before they were placed on the torture-chamber trunks where dragons awaited them. If all had been above board, it would have been clear that gray moths and black moths are nothing other than variants of one and the same species—siblings that differ because of a single mutation or perhaps, as someone suspects, because of the environment, like Weismann's *Vanessa* butterflies.

The peppered moth is the most frequently cited example of "evolution taking place before our eyes." On the contrary, evolution is not something that happens under our noses but is a long, long process (and a rare one). It is like the extremely slow movement of certain "fixed" stars—fixed vis-à-vis the rest of the firmament—or the rotation of a galaxy; it is not apparent to a person observing the night sky, unless he is mistaking meteors on the mid-August night of Saint Laurence's Day for real stars.

Natural Selection, which indeed occurs in nature (as Bishop Wilberforce, too, was perfectly aware), mainly has the effect of maintaining equilibrium and stability. It eliminates all those that dare to depart from type—the eccentrics and the adventurers and the marginal sort. It is ever adjusting populations, but it does so in each case by bringing them back to the norm. We read in the textbooks that, when environmental conditions change, the selection process may produce a shift in a population's mean values, by a process known as adaptation. If the climate turns very cold, the cold-adapted beings are favored relative to others; if it becomes windy, the wind blows away those that are most exposed; if an illness breaks out, those in questionable health will be lost. But all these artful guiles serve their purpose only until the clouds blow away. The species, in fact, is an organic entity, a typical form, which may deviate only to return to the furrow of its destiny; it may wander from the band only to find its proper place by returning to the gang.

Everything that disassembles, upsets proportions or becomes distorted in any way is sooner or later brought back to type. There has been a tendency to confuse fleeting adjustments with grand destinies, minor shrewdness with signs of the times.

It is true that species may lose something on the way—the mole its eyes, say, and the succulent plant its leaves, never to recover them again. But here we are dealing with unhappy, mutilated species, at the margins of their area of distribution—the extreme and the specialized. These are species with no future; they are not pioneers, but prisoners in nature's penitentiary.

The problem reduces to this: Either the type is a solid, stable kind of reality, able to defend itself through oscillations, retreats, and advances; or else the type doesn't exist at all—only the mean exists, as the statisticians will have it. If the mean changes, there is no reason why it should change back again. Unless there is a model to adhere to, everything tends to dissolve or disperse in a myriad of practical solutions. "While the epochs go forward without rest," says Manlio Sgalambro, "truth stays motionless, and real strength lies in staying motionless with it."

Living nature does not play the stock exchange; still less does it play roulette. There can be no denying that chance has its part to play, but to claim that it is the governing reality, allied with opportunism here and there, is without foundation. Chance could be invoked in explaining the vagaries of species, their restless wandering, their gradual conversion one into another. But none of these things exists in nature's repertory.

Species maintain stability, in some cases for millions of years, despite the assaults of mutation and the pressures of selection, and then they disappear. The opossum (*Didelphis marsupialis*) is found throughout the American continent, where it climbs trees and is persecuted for the damage it wreaks on poultry. The female gives birth up to three times a year, each time with a litter of ten to eighteen. Fossil opossums from the Cretaceous (about 100 million years ago) appear to be no different from the chicken predator of today. Despite its great prolificacy and the extreme range of environments in which it is found, the species has remained faithful to itself. Another interesting case is that of the *Lingula*, a bivalve shellfish in the brachiopod phylum. Several species have remained virtually unchanged since the beginnings of multicellular animals about 550 million years ago. How could this have happened without some mechanism of stability—indeed something harder to imagine than the source of variability? Why trouble our brains to find an explanation for something that doesn't happen anyway? Why dismiss haughtily the real mark of life, namely stability?

The Earth's crust bears a greater range of variability than the organisms that make their home on it and from which they wing aloft. The continents have drifted, collided and interpenetrated with each other while species have conserved their features. As Thorpe put it: "The wagtail (*Motacilla*) in the garden was there before the Himalayan mountains were formed."

CHAPTER V

LEARNING THE INNATE SONG

Over the last two centuries people have had an idea—a sort of practical rule of thumb—that every part of an animal or a plant is fashioned for a purpose: wings to sustain birds in flight, and floral corollas to attract pillaging insects. The idea is quite clever, and it explains how the parts have been maintained all along and in good condition. Wingless swallows or flowerless rose gardens would have little chance of survival and would disappear from the scene, if not from nature's severity then from their own sadness. This idea is less successful, however, in explaining how in the course of their history birds obtained wings and rose gardens acquired the flowers that give them their name.

A swallow cannot have obtained its wings as though ordering a tuxedo from a tailor. A swallow without wings would not be a swallow; it would not be in any condition to drag itself out to get fitted at the emporium of reality. So some people supposed that, a bit at a time, as if in a lease-to-buy arrangement, the swallow acquired its splendid black-arrow attire starting out as a fledgling chick, then became sparrow-like, and finally metamorphosed into that aerial silhouette we recognize as the harbinger of spring. But it is difficult to imagine how it managed in the meantime as a half-swallow, and why it should have labored away at a journey with no destination in view; for, to be sure, it never saw the goal of its laborious pilgrimage. Did anybody ever see a half-swallow, or has some paleontologist unearthed a fossil of a half-bat at some supposed intermediate stage of design? Wherever they appear in the fossil record, swallows and bats are completely formed, and before them there is nothing presaging them or approaching them in likeness.

Richard Dawkins has tried to reconstruct the historical process with the help of a computer. Within a reasonable period of time he succeeded in evoking on his monitor the outline of a fly, or a verse of Shakespeare, as the outcome of a series of blind, successive "mutations" starting with a vague sketch or a line of randomly chosen letters. Let us take the line of poetry. Randomly varying the individual letters of the alphabet in the initial nonsense makes, then unmakes what it has made; it approaches and then retreats. It would take billions and billions of years to compose the desired verse or, indeed, any verse. So much for Dawkins's own calculations. Cicero was aware of the same problem, two millennia earlier, when he wrote that it would never be possible to write a verse of the poet Ennius by tossing letters randomly on the ground.

So how did Dawkins manage it? What he did was to load a verse of Shakespeare in the computer's memory, then establish the rule that any letter randomly appearing at the right place should stay put. One after the other the letters settled in their appointed places, and presto! There on the monitor was that verse of Shakespeare's. Dawkins describes all this in his book *The Blind Watchmaker*, implying that living forms come about without any plan, by the play of chance. In reality he demonstrated just the opposite: Unless there is a plan, a pre-established design, nothing—nothing at all!—can come into existence. The verse that appeared so miraculously on the monitor was there already, hidden away, beckoning the letters to take up their places in the ranks according to Dawkins's instructions.

Does this mean that our swallow or bat was there already, in nature's catalog, before Being, incrementally or all of a sudden, took up residence in Form? René Thom, the mathematician of Catastrophe Theory, offers a clear statement of the concept, comparing living structures to the figures of geometry and saying that there will be a limited number of these. Each archetypal form, when caught up in the wave of existence, becomes reality.

The living form tends to express its own identity or bear witness to its own nature. It does this by taking on configurations, exhibiting designs, uttering song or giving off perfume—none of these having much to do with survival, with utility, or with vital functions. Karen Blixen, the great Danish writer (*Out of Africa, Ehrengard*), was enchanted by chestnut blossoms as straight as candles, by lilacs bursting out on every branch, by laburnum flowers hanging like golden icicles, by hawthorn blossoms in white- and rose-colored abundance. She could not conceive that such infinite variety should be necessary for the economy of nature. She sees it, rather, as the manifestation of a universal spirit of joy that is "incapable of holding back its playful torrents of happiness." If we attempt an explanation for the variety of all forms by referring them solely to the parameter of survival, or of utility, we are denying nature's peculiarities their specific *raison d'être* and handing them over altogether to some "reason" that is external and extraneous to them. We are putting them all in a spreadsheet where one form is worth no more than another and none is valued for itself. If we do this, we destroy singularities and physiognomies and put in their place something like a tax identification number, whose only function is to ensure that no one goes tax-free. So, take arms, you species, families and orders, and insist on having your own *raison d'être*, your own soul. Refuse to give account of yourself to the taxman!

The problem of why species are so greatly different from one another is not a problem of adaptation or opportunism. The differences are there in far greater numbers than are functionally necessary; they are rather like stubs in the checkbook of belonging, like outbursts of life. Adolf Portmann remarked that the higher animals—tiger, deer, human—present themselves in their very bearing as beings conscious of their role in nature. The higher the organization, the more the expression of form becomes concentrated in the head, the polestar of the organism. The spitting tiger's visage is a masterpiece of the power of expression, the dentition exalted by projecting canines, the threatening eye surrounded by savage lines, the whiskers outstretched like luminous traces, the white coat framing the expression of ferocity, the diabolical ears folded backward. The stripes follow the contours of the body as though they were brush-strokes emphasizing its massive musculature and its impending leap. The tiger is not something adapted to the jungle, nor does it emerge from the jungle. It storms into it; the tiger makes the jungle. What nobility, too, in the muzzle of the horse, its eyes, ears, nostrils; and what dignity in its gait.

In lower mammals such as squirrels the body has a banded pattern that divides up the trunk, while the snout is inexpressive and without markings. In fish the colors can be bright and resplendent, even among species that never see the light of upper regions, but their patterning bears no relationship to internal structure; the colors just seem to be put there like paints on an artist's palette.

Mollusk shells are past masters of form, with their torsions and decorations, but they enjoy few admirers in the depths of the seas. They are like works of art from the sculptor's or painter's hand, masterpieces on which the artist labors with great skill but complete disinterest, until they sink down to the depths of the ocean, there to be seen no more. The spiral shapes of shells, D'Arcy Thompson observed, are exercises in mathematics, products of a nature that exhibits only the reflection of forms contemplated by geometry. The patterns are the work of delicate abstract painters who have no art galleries and no art market. The shells of *Voluta* are "painted" in alternate light and dark bands. The edge of the mantle, growing like an advancing wave, leaves colored secretions in its wake. If the secretion is limited to certain segments of the margin, the shell acquires a meridian-like coat of bands perpendicular to the edges of the mantle. If the secretions vary in time, the bands alternate parallel to the margin, which now paints, now rests. If the two patterns are superimposed, the result is a white net-like pattern that separates series of tiny, dark squares in ranks and columns. And all this for the joy of the geometrician or of some god forever doing geometry.

What the animals have to say, and to tell us, has little to do with comparative anatomy. Anatomy instructs us how the whale and the cat are made in the same way, apart from exaggerations and atrophies in their various parts. The latter are differences for which we have adaptive explanations. Comparative anatomy is the science of homologizing parts of animals. A bird and an elephant have the same vertebrae, the same ribs, the same number of bones in the feet. A bird's wing is simply an atrophied foreleg; a whale fin is a comb-patterned paw.

Physiognomic morphology, which looks for the overall meaning of animals as animals, tells us what the animal has to say. It is a science that does not enjoy much repute because it is the stomping ground of the anthroposophists and is contaminated with poetry. Taking our cue from Herman Poppelbaum's descriptions, let us try to compare the bird and the mammal.

The quadruped mammal, with its fur, its scent and its secretion, tends to melt into its surroundings. It takes the environment into its viscera and fertilizes the environment with its organic feces. It forms its offspring inside its body and nourishes it with its milk. It is guided by its own heavy head, which noses its way in and out, with its powers of smell and taste in a sort of embrace with mother earth. Its colors are gray-brown like the earth itself. It trembles, whimpers or cries out. The bird has its trunk enclosed in a bony basket that contains the abdomen in its vault. Its tiny head is attached to the trunk rather like a tentacle, detached like a periscope. This is an instrument for pecking, for nibbling and for nest-building. The feet are dried up, stunted and mechanized. The offspring is made to grow outside the body, enclosed in an inaccessible shell. The feathers form a horny armor that renders the creature impermeable to the environment. A bird without its feathers is unrecognizable and appears reduced to "a little bundle of misery." The bird is formed by its feathers, which mark out its contours and express its colored splendor. Erect on its two feet, it wings away free from the earth; its kingdom is the air, the atmosphere, the far-off. In its lungs and its air sacs it takes in lots of sky, making it fragile and light. The thrush thinks with its lungs, said Rudolf Steiner. The glory of the bird lies in the wings that turn the tiny bones of its "arms" into enormous fan-like extensions made up of feathers that are splendid and dead. With these, the bird wings its way aloft and sketches in the heavens the trajectory of its flight.

And sings.

The *Silviidae* are minute, gray-brown members of the sparrow family, the best known of them being the blackcap. They excel in their youthful song, beginning their do-re-mi exercises in their second week of life and completing their joyous repertoire after a month. Bird song is unique to each species of *Silviidae,* and no drill or learning is required. Even if the bird has been kept isolated from all sound, even if its egg has been kept in silence, its song bursts forth spontaneously when its moment comes—and all the experts agree that no "adaptive" function is at work. In some species the song comes out even in individuals deprived of hearing—individuals who perform their Ninth Symphony without ever listening to it. An observation of Konrad Lorenz corroborates this finding. The bluethroat (*Cyanosylvia svecica*), he tells us, gives out a harmonious, complex song, a rare musical delight; it makes music when it is tranquil and relaxed, sitting in its bush "poetizing for itself." When the adult is busy singing for some purpose or other—to defend its territory from a rival or to attract the female—the delicacy disappears and the bird repeats monotonously only the strongest "passages," now graceless, but with decided intent. Bird song, Lorenz tells us, is the birds' innate expression, a burst of "self-presentation," to use Adolf Portmann's term. In no sense is it derived from the bird's necessities, since these, if anything, have a degrading effect on it. For its adaptive purposes, the adult bluethroat selects from its vast repertoire only a few strident passages, and in the struggle for survival it expresses far less than what it has at its disposal.

With canaries, the story, it seems, is different. Canaries have to learn their song, memorizing the sounds that they will later emit. This does not mean that their song is not innate, but rather that it remains concealed and they produce it only when it is evoked. They recall a silent song, a song they hold inside themselves. Other birds, such as the nightingale, even learn and reproduce the bird song of certain other species, yet they return each time to their ancestral one. Learning expresses the unexpressed greatness of the innate.

The innate endowment is there not only in the spirals of their DNA, but it is also at home in all those constraints and cages to which the developing bird conforms. It is in the outpourings of the morphogenetic field that burst forth in the splendor of feathers and in the festive melodies of song.

I realize that everything I am saying here is permeated with one conviction, namely, that the innate must be good, that its revelation must be something consoling and joyful, like the music of the lark. And my feelings contrast with the peevishness of evolutionism, according to which "congenital" immediately suggests a taint—an embryo aborted, or some primordial ugliness; and according to which the goal of life and progress is deliverance from the obligations of a painful contract with the past.

I myself feel overcome by the innate that reveals itself as something sublime and unique, peerless and absolutely beyond the reach of my judgment. The beauty of the world is unbounded and meticulous compared to the mindless idiocy of cyclones and squalls, or compared to the grammatical mistakes in the recital of the divine comedy of Being. Should we believe that outbursts of chaos and threads from the past have built us and our promise of the future? The budding flower of the world is a cathedral of cathedrals, and it remains to us to bend our knee and say "Domine, non sum dignus."

CHAPTER VI

WHY IS A FLY NOT A HORSE?

The scientist enjoys a privilege denied the theologian. To any question, even one central to his theories, he may reply "I'm sorry but I do not know." This is the only honest answer to the question posed by the title of this chapter. We are fully aware of what makes a flower red rather than white, what it is that prevents a dwarf from growing taller, or what goes wrong in a paraplegic or a thalassemic. But the mystery of species eludes us, and we have made no progress beyond what we have long known, namely, that a kitty is born a kitty because its mother was a she-cat that mated with a tom, and that a fly emerges as a fly larva from a fly egg.

We do not know the answer in the terms in which science has sought to provide it—the same terms as those which have provided a satisfactory explanation of the differences within the species, namely chromosomes, genes, DNA. If we want to solve the problem underlying every origin of species in molecular terms, we have to admit that for the moment the answer is not forthcoming. And there is no answer, despite our having repeatedly fancied we had arrived at the last door, beyond which the answer would have been revealed. Even though we have learned to decipher the hieroglyphic that enables us to read the latest inscription, it is an inscription revealing only generalities.

By adopting the method followed by Mendel in his studies on the garden pea, it is possible to establish whether a morphological or functional difference is due to a difference in a gene. Because of a gene, a strain of pea with red flowers is different from a strain having white flowers. The same is true of the fruit fly with red eyes as opposed to one with white eyes. Hemophilia in man is associated with a gene located on the X (sex) chromosome, and thalassemia with a gene that synthesizes beta-hemoglobin. But genes for Cat and genes for Fly have defied discovery; and the same can be said of the genes for eyes and those for ears.

At one point it was thought that cat-ness was a diffuse prerogative of cat genes and fly-ness of fly genes. With the advent of comparative protein analysis it seemed that the demonstration was at last within our grasp. Proteins are the direct (or almost direct) products of genes. "One-gene-one-protein" was the exciting implication of biochemical research centered mainly on bacteria and molds. Genes and proteins are both polymers—molecules made up of subunits in series. The beads in the gene necklace are nucleotides and the beads in the protein necklace are amino acids. Each nucleotide triplet "codes" for one amino acid, so a gene with nine hundred nucleotides manufactures a protein having three hundred amino acids. The nucleotides are of four types only, whereas there are twenty types of amino acids. The algorithm that relates the 64 nucleotide triplets with the 20 individual amino acids is referred to as the genetic code.

The somewhat infelicitous expression "the genetic code" of, say, Raphael or Shakespeare has gained currency, like "your code" and "my code," or "cat code" and "fly code." The expression is mistaken because the genetic code is universal and Raphael had the same code as his housefly and our fly—the identical code book for translating his genes into proteins. As a matter of fact, researchers who cracked the genetic code immediately realized that it was universal. Nor—as if they knew this already—were they greatly surprised. For the moment, then, let this be clear: It is not the genetic code that accounts for the difference between the cat and the fly or between a virus and a whale.

Actually, when a person uses the term "code" improperly, he is referring to the genetic endowment or—in literary terms—to the genetic "text," which is decoded into the protein endowment. The latter indeed could conceivably explain the difference between our master painter and his irritating house-fly.

And so, let us restate the question: Do differences in the genetic text or in its resulting proteins explain why our animals differ from one another? Nowadays we know the text of many widely encountered proteins, and we can provide an answer. Let us see how things worked out.

It all started with protein analysis. Many proteins are ubiquitous—i. e., they are to be found in all species. Strikingly enough, bacteria, animals and plants have 80% of their enzymes in common. The earliest protein analysts selected an enzyme about 100 amino acids long that could be found in all organisms: cytochrome c. This was analyzed, letter by letter, in a vast number of species. In humans and chimpanzees this molecule differs by a single amino acid residue; the difference is at position 66, which in humans is isoleucine and in apes threonine. The monkey and the horse differ in 14 amino acids, the horse and the rattlesnake in 21, and the snake and the shark in 24. There are 25 differences between the tuna fish and the fly (including four amino acids that the fly has in excess over those of the vertebrates). Within a certain approximation one might have concluded that the more distantly related two species are the more they differ in the molecular composition of their proteins. Having reached this stage, some people imagined that the feat had been accomplished, and that at last we understood at the molecular level what made species different. Since behind every amino acid change is a gene mutation, it seemed that the latter offered the ultimate explanation of what made a cat a cat and a fly a fly.

Yet the explanation failed. It gradually became clear that the molecular differences were irrelevant for protein function. The cytochrome c from all species—man, fly, spinach—was functionally identical, and the mutations that were thought to have accounted for the molecular differences were termed "neutral." Spinach cytochrome c did its job just as well as human cytochrome c.

"The more one approaches the molecular level in the study of living beings," R. E. Dickerson wrote in 1972, "the more similar they appear, and the less important the differences between, for instance, a clam and a horse become." From a biochemical standpoint the horse and the horse-fly are essentially the same. The biochemistry of living things is substantially universal. It has remained constant over geological eras and within species that are morphologically entirely dissimilar. This has been the great message of molecular biology. In 1977 François Jacob, a founding father of biochemical genetics, wrote: "Biochemical changes do not seem to be a main driving force in the diversification of living organisms.... It is not biochemical novelty that generated diversification of organisms.... What distinguishes a butterfly from a lion, a hen from a fly, or a worm from a whale is much less a difference in chemical constituents than in the organization and the distribution of these constituents."

It is the observable differences among individuals within a species, between you and me, that are biochemical. Each species has its spectrum of variants, rather like a nosological registry of potential deviations that correspond to the more frequently encountered mutations in the genetic endowment of a species. This is most readily detectable in domesticated species, humans included, where variations are retained that environmental pressure tends to eliminate in wild species. Russian geneticist Nicolay Vavilov, in his monumental work on cereals, noticed that the various graminaceae species—wheat, barley, millet, oats, and rye—display the same spectrum of variability. All have ears with or without awns; all are smooth or rough, early or late; all have dark colored seeds or light colored ones; and so on. To these patterns of variation he gave the name "parallel series," and based on them he was able to predict variations unknown at the time and subsequently to discover these in his botanical explorations.

Whereas almost all cereal species had winter or spring varieties, hard wheat (*Triticum durum*) has a winter variety only. Vavilov searched for a spring durum and found it at last, in 1918, in an isolated region of northern Iran. He spoke of "internal constraints," or "laws of form," which are intrinsic to species and common to related species.

Related species share morphological varieties, that is, the same pattern of mutations. This is indicative of their genetic similarity but does not say what makes one different from another. We can illustrate the phenomenon of parallel series by comparing two sets of chessmen, one black, one white. The two formations have the same shape variants—king, queen, knight, castle, bishop, pawn—and they differ in one shapeless trait—color. All this eludes the methods of the geneticists, who have to cross different forms. We can say that the two chess armies differ much less from one another than within themselves. And it is precisely the singular quality distinguishing species that eludes our analysis. The members of one species can interbreed, yet they generate no fertile progeny with another species. Let us say that one could breed a black king and a black queen, but not a black chess piece with a white chess piece.

Among the mutations that alter the forms of animals some are dramatic and produce veritable upheavals in morphology. Such mutations were first observed in the fruit fly (*Drosophila*), where parts were displaced, lost, or duplicated. There are *Drosophila* with legs instead of antennae, *Drosophila* with a double thorax (and thus with four wings), and *Drosophila* entirely without eyes. Ah-ha! Have we here discovered the genes responsible for overall organization—for determining, in other words, respective positions within the body? In the last few years it has become clear that the genes in question are found all together in an aligned series of ten—the Hox cluster, as it has come to be called. The cluster looks like a genetic miniature of the fruit fly: The first gene governs the front segment, and the others, in line, are responsible for controlling succeeding segments.

An unexpected and extraordinary discovery indeed.

The greatest surprise of all came in 1989, when the same cluster of genes (i.e., the Hox cluster) was discovered in mice, then lancelets, frogs, leeches and worms—in short, in every kind of animal, humans included. In vertebrates, the cluster was duplicated four or five times. In the different animal types, each gene in the cluster is responsible for the development of a given region in the embryo, from the head to the tail, though giving rise—obviously—to different expressions in different animals. The gene governing the cephalic spiracles (respiratory pores) in the fruit fly is expressed in the head of the mouse. The same gene that governs the tail of the mouse governs the posterior extremity of the grasshopper or amphibian. All animals have the same cluster of genes dictating the order in their body regions. The cluster, therefore, was universal and of great antiquity, since it belonged to the progenitor of all animals and had been preserved intact throughout the development of the fauna for 500 million years. The genes decreeing the general organization of a mammal are the same as those decreeing the organization of an insect.

If we transfer the gene responsible for the occurrence of the eye in a cat to the egg of a blind fly that lacks its eye-forming gene, the fly develops its normal red-faceted eyes despite the fact that the gene came from a cat with round blue eyes.

What is it, then, that makes a fly a fly and a cat a cat? The genes responsible for the ordering of the body are universal in animals (just as the genetic code is universal in living things). Thus we are still in quest of that certain something that makes the elusive difference. The view is gaining ever wider support that this something is not to be found in the innermost molecular heart of the cell, but perhaps in some vague "field" that unfolds to the point of being the very form of a fly or a cat.

And one suspects that these final forms are not just the outcome of brushing genetic paint on the outline of this "field," but that they themselves have something important to say about the construction of living beings.

The crystalline lens of the eye is produced inside the frontal epidermis when the optical vesicle, a protrusion from the embryonic brain, comes to rest on the epidermis. The lens is then framed by the colored iris, derived from a totally different tissue. If a salamander has its crystalline lens removed, the latter regenerates—but not from the forward-facing epidermis where it originated. It forms from cells on the edge of the iris epithelium, which lose their color, start dividing, and become transparent crystalline lens. Driesch gave the name "equifinality" to this capacity for achieving the same morphological result via different routes.

A final form, one might say, draws development unto itself, making it occupy its appointed space. If so, not only the salamander's eye but also the entire animal—indeed, every form in nature—seems to strive toward a receptacle prepared for it, toward a pre-arranged landscape. "Every proper form," according to René Thom, "aspires to existence and attracts the wave fronts of beings."

This way of thinking finds similar expression in the "convergent wave" theory of Luigi Fantappié. A stone falling into a pond creates a local turbulence that expands in concentric and diverging circles that gradually become thinner as they move outwards and eventually disappear. If we could reverse time's arrow on the still pond, a thin, wide, circular wave would form and move toward the center, its diameter becoming smaller and its crests becoming higher, to be followed in turn by other waves converging, like it, from nothingness. Finally, as the wavelets vacate the field and become concentrated at the center, a turbulence would erupt that spits out the stone. Fantappié (1941) called this process "syntropic" to contrast it to the "entropic" decay of the physical world. About the same time Schrödinger (1944) advanced a similar concept by the name of "negentropy." If we transfer this to the world of living things, the converging wave is like a frog that becomes ever smaller and changes into a tadpole, a gastrula and finally an egg. This, of course, never happens; nevertheless, if we invert the order of time we can sketch a future that beckons the present and accompanies it along the appropriate—equifinal—pathways in order to actualize it. No such experiment is feasible; but the story can be told and, for all its implausibility, it offers a strange inverted logic that is capable of describing in mathematical terms how final forms draw reality unto themselves.

In so-called "primary induction," the blastopore lip of the frog corresponds to the stone falling into the pond. It produces folds and cavities that eventually become a frog. But we know that any organic —even dead or inorganic—particle can replace the lip. The latter is not the cause of the process, which is pre-arranged and "self-organized" and only awaits a signal to respond to the call—opening itself up to the farthermost wave that marks its confines. In this crystalline lake, destiny—the innate—does not impose itself as coming from the past but as waiting in the future, as the ultimate wave that implodes in the direction of the present. The "final wave"—form—does not produce the events that make up development and life: It adopts them and calibrates them so they will reach their goal. The "syntropic" world is a parallel universe, which circulates backwards in time. But has not destiny always been situated in the future? And haven't those who assign "fate" to the genes actually inverted the meaning of the word—perhaps so they can manipulate life at their will?

CHAPTER VII

THE ETERNAL CHILD

The Pius IV Pavilion in the Vatican Gardens, a freestanding villa surrounded by trees and lawns suggestive of lost paradises, is the venue for the meetings of the Pontifical Academy of Sciences. In 1982 I was invited, because of some supposed merits of mine, to take part in an international workshop on human origins. The actual title of the meeting was "Recent Advances in the Evolution of Primates." Contributing to the workshop were the most outstanding figures in human paleontology and molecular biology—experts who (unlike myself) had long been studying humans and apes and who already agreed on the general lines of how to deal with the subject.

The primate paleontologists had to take into account an announcement made a few years earlier by the molecular biologists. The latter had just begun looking into human molecules and those of our cousins, the apes. In the light of certain comparisons and calculations, which it is not my purpose here to illustrate, they advanced the hypothesis that humans and apes had taken their separate directions on the evolutionary tree not more than 1.3 million years ago. This was hard for the paleontologists to swallow, because they had been unearthing human fossils at depths dated at three million years and had found traces of hominids of erect posture (*Australopithecines*) five to six million years old. Since it was received wisdom that humans were descended from the apes, how could there have been humans and hominids antedating by millions of years the fateful split that supposedly gave rise to them? At the time, the spotlight was still on an Asiatic specimen (*Ramapithecus*) that most agreed was a hominid, belonging somewhere on the line leading from apes to humans, yet the impertinent *Ramapithecus* had lived fifteen million years before that supposed split.

Anyone looking at science from outside thinks that scientists are at pains to come up with theories that fit their data. Yet the contrary is often the case (and here we have one example), with scientists striving to force the facts into some preconceived theory. Man the progeny of the ape was gospel for the evolutionists—a symbol of the new world unquestioned since the dismissal of Adam and featured on the covers of our school textbooks. Among the experts no one would have dared openly to defend this fateful descent, but everyone took it as tacitly agreed—as if there were a moral obligation to do so, a categorical imperative never to question it.

In this way a scandalous compromise was reached. The unfortunate *Ramapithecus* was ejected from the hominid branches of our tree and relegated to another tree in the forest. The molecular biologists agreed to push back the split between apes and hominids to seven million years ago, and the paleontologists promised not to place the hominids at a date earlier than the agreed-upon seven million years. The compromise entailed a highly dangerous concession from the biochemists, who thereby had to admit that the so-called "molecular clock" that marked the accumulations of mutations over time had ticked away at a slower rate in the hominid line.

At the Pius IV Pavilion no mention was made of fossil apes, chimpanzees, gorillas or orangutans (or their relatives in the ascending line), though these would have rightly played an important role in the pattern then taking shape. Only some time later did I realize why. There are no ape fossils. There are some partly fossilized specimens, but these date back a hundred thousand years or little more. Something missing is always less embarrassing than something there; but the total absence of fossil apes (technically "pongids") was simply inexcusable. Our beastly ancestors were unrepresented in our gallery of family portraits.

To be sure, if it had been a question, not of His Majesty Man, but of birds or frogs, dating the arrival of an animal without fossils to a later period than that of the animal whose fossils had been found at three or four million year old depths would have been the most obvious operation for our paleontologists. But here we had a mess that concerned us humans, and dirty linen is better washed in private. I have no wish here to delve into paleontology and the story of hominid crania and bones from the past few million years—*Australopithecus, Homo habilis, Homo erectus,* and *Homo sapiens*—because I do not feel sufficiently expert concerning what are, not inappropriately, called "bones of contention." In any case, little was said about these at that Vatican workshop.

Other facts, more interesting and less controversial, emerged from the comparison of anatomies among extant species and their molecular endowments.

By observing the traits of related species, anatomists distinguish between "primitive" and "derived" characters. To put it simply, a trait is "original/primitive" whenever it is close to conformations typical of the order in question and similar to that of its most ancient representatives and, at the same time, close to whatever conformation in their embryos is common to all species in a given order. On the other hand, a trait is "derived" when it has become transformed in comparison with archetypes, is functionally differentiated, and is adaptively specialized.

Related species belonging to the same family have been categorized by some systematists (such as Daniele Rosa in the early twentieth century and Eugene Balon in the second half of the same century) with respect to their infantile-senile character. This has nothing to do with paleontological order but concerns only development, which proceeds slowly in the infantile/primitive species and faster in the senile/derived ones. An infantile species is more active, more beautiful, and more spirited compared with its senile siblings. Take the infantile horse and the senile pack animal. The horse is irascible, statuesque, full of fire; the donkey is stubborn, mean, indolent. The latter is usually rated more intelligent than the former—the sign of modern outlook that prefers the practical to the inspired, the indifferent to the passionate.

Man is more juvenile than the ape, and woman more than her mate. If we were to continue with these assignments, which are becoming something of a joke, I would say that the plant is more senile (more rigid, colder) than the animal—except for its flowers, in which all its vain youthfulness explodes, passionate and free, and in the rose above all, the queen of plants, remaining ever in the bud.

According to these definitions, Man displays a splendid repertoire of "primitive/infantile" conditions, whereas his tailless cousins exhibit in those same traits conditions that are unmistakably "derived/senile."

The human cranium lacks crests and prominent brow arches; it lacks the exaggerated muzzle of the chimpanzee; it lacks protruding canines. All this "lack"—this geometrical rotundity (apart from the nose)—is a primitive condition. We find it in the oldest fossil primates; we find it in the embryos and young of monkeys, which take on bestial aspects as they grow older—and older they quickly become. I once saw, in a zoological museum in Salzburg, a panel displaying glowering gorilla skulls. Among them was one tiny delicately featured specimen—human, one could have thought—which was the cranium of a newborn gorilla.

The occipital foramen is located lower down in the human skull, articulated on a vertical neck in a central, fetal position. The same occurs in the ape fetus, but as the ape develops the foramen recedes and the brain remains as if fenced in between neck and jaws. What nobility there is in the erect, but not rigid, column of our species, compared with the senile, humped backbone of our knuckle-walking brethren!

And take the human hand. The hand! How open and suggestive of fair play is the human hand, with its fine fanning configuration, an original/primitive architectural model. Compared with the human, all other mammalian hands are deformed and sacrificed to specialization.

The ape hand is elongated and hooked, the thumb stunted; the lion's "hand" is contracted and fitted with claws; the bovine hand is reduced to two fingers and the horse's to a single finger; while the cetacean's is an ankylosed fin.

And our great head, too, of which we are so proud, is a fetal trait. It is relatively larger in the newborn and the baby than in us thinkers, and in the prosimian compared to the anthropomorphic apes. What it all boils down to is this: The human form is the most original, archetypal, and primitive of all the mammals. It is the form of the child, the dawn, the exemplar. We could say that it is the most primordial of all mammalian forms, provided the term "primordial" is not used to connote the brutality of gnostic beginnings, or "primitive" to suggest coarseness and savagery.

So arise, people, especially you ladies and young ones. Lift up your gaze, stand up straight and stretch out your arms. You are models of the beautiful, the original, the ideal. You were here at the beginning, and in your bodies you give witness to the primordial, unchanged antiquity of being.

The embryos of all vertebrates resemble one another midway through their early development. At this stage they look like tiny croissants, each with a huge head between two staring eyes, and a pointed tail. As they grow, they diversify into the various classes (Birds, Reptiles, Mammals), orders (carnivores, cetaceans, primates), and families (lemurs, pongids, hominids), etc.

In his novel *The Once and Future King*, T. H. White tells a story that has been taken over by Stephen Jay Gould. It goes like this: God created a series of embryos all like each other and then made them parade before His throne. To each one He asked what beak, what claws, what fangs it would like to have when it grew up. Each embryo duly made its request, until it came to the turn of the human embryo. The human embryo said this to the Lord: "If I can have my way, I'll stay as I am.... I'll remain a defenseless embryo all my life." God was pleased with this and said: "You shall be like an embryo until your dying day, eternally a child.

You will remain omnipotential, in Our image and likeness; and you will be able to understand some of Our sorrows and experience some of Our joys…"

Eternally a child, then. This is the destiny of the least specialized of beings, not only in the form of the body but above all in spirit—called, unequipped though he is, to face the world about him. He must ever be learning something and discovering one or another of the hundred stories written within himself; ever undecided, ever at a crossroads from which the paths lead on to sanctity or to perdition, to the poetical or to the banal. Hector Bianciotti says that every man contains within himself partial men that know not one another; we are born "plural," we die a single person—or a nobody. Among all beings, Man knows least who he is and what destiny awaits him. Embryo of all the species, he can become lion or lamb, eagle or chicken; remain a worm or turn into a butterfly. He is a lost bird seeking to learn its bird song. Faced with such a range of choices, his true resource lies in conserving his initial naivete, in leaving inside himself something ever unexplored, in emulating Peter Pan, the little monarch of Kensington Gardens, the eternal child.

The interesting thing about our story is that it was not an embryologist or an anatomist who called us Peter Pans, but a molecular biologist, Alan R. Templeton. Templeton compared corresponding molecules found in different extant species and was able to time the rate at which their transcripts became modified over geological eras. At first it was thought that the rates were the same for all groups of living beings, and the expression "molecular clock" came into use—a clock measuring so many mutations per so many "letters" per unit of time. In the ubiquitous molecule cytochrome c, a mutation clicked by every twenty million years. Other macromolecules were amenable to even more precise timing. All went well enough until Morris Goodman began using the universal clock to investigate the human line. It was precisely here that things did not work out. Rhythms slowed down; the clock went awry and the molecular data conflicted with paleontological findings. This is the disagreement we spoke about at the beginning of this chapter.

Once the universal "molecular clock" was shelved, biochemists ceased to question the (in any case dubious) datings proposed by pale-ontologists. Having taken these datings as valid they could now devote themselves to a somewhat less ambitious occupation—measuring the ticks in a gallery of unsynchronized clocks. The genetic text does not "decay" at the same rate in all species. Unlike the impassive Carbon 14, which tells us the age of some bones and fossils, the molecular "time" is intertwined with living things.

With paleontological dating of the histories of humans and horses, which began together and continued in parallel, it has become clear that the steed's molecules have "decayed" seven times faster than those of its rider. From the moment that the human line became detached from that of the chimpanzee, hominids have clocked 13 "misprints" in mitochondrial DNA compared to 34 in the chimpanzee. This means that, from the chemical standpoint, humans have changed much less than chimpanzees when compared with their common ancestor. In strict terms this shows that humans have evolved much less than African apes, implying that the common ancestor was, if anything, more human than ape; and that the human line has remained, as it were, more baby-like than its quadrumanous cousins. It was in light of these data that Templeton concluded: "Man is the Peter Pan of the primate world—the boy who wouldn't grow up." A "molecular" Peter Pan.

During the same years, cytologists compared the fine structure of human and ape chromosomes and came to the same disconcerting conclusion: The chromosomes of the mysterious common ancestor were similar to Man's. Man was revealed under the microscope to be a "chromosomal" Peter Pan in comparison to the ape and the gorilla—the latter's ideogram showing pronounced signs of the wear and tear of life.

Darwin's conjecture that the common ancestor of Man and the apes was substantially ape, and that this ape had remained largely what it had been from the start, thereby collapsed. The truth was that Man had remained what he had always been. At the parting of ways the molecules and chromosomes of the human being were already there.

It is indeed surprising that we have been led to this conclusion by the molecular biologists and cytologists—the very people who were probing into the depths of the cell's heart that had seemed so unlikely to provide information about the causes of differentiation and the evolution of species.

How is it that humans should have come to a chemical halt? Had they reached perfection? Were they tired, perhaps? Had they developed effective systems for warding off mutations? Or was it a case of social structures favoring biochemical stasis? None of these questions has received an answer; but molecular revelations after all were capable of bringing about that upheaval in the accepted view of our origin—revelations that the most highly qualified research in anatomy and paleontology had failed to elicit.

Molecular revolutions do have something negative about them: They fail to touch the collective imagination. They are difficult to understand and explain. They remain in the exotic language of specialists, which fails to communicate not only to the public but also to scientists in the next room.

Thanks, therefore, to molecules! And even if they have added mystery to mystery, they deserve our thanks for having restored to Man the youth and the dawn that the theory of evolution had taken away from him—without proof, and simply on the presumption that the ugly and the unrefined must come first and the gentle and the graceful must be their descendants. What an absurd prejudice, that first there was old age and then childhood blossomed from it!

From what we have been saying, it is clear that humans are distinguished from the apes and—who knows?—distinguished from all other mammals, by having been exempted from evolution, by having remained what they were in times now lost beyond recall, castaways unscathed by the storms that have denied their fellow animals the erect posture, or have made them grow fur or fleece and arm themselves with fangs and claws. But how are we to distinguish the ancient from the modern? Were not the ancients modern in their day? The modern is the offspring of history, something which becomes worn in experience, something that is anxious to grow up; whereas what is ancient is the offspring of something perennial and immemorial. The ancients were never modern, for they harbored within themselves the sign of the absolute, a closeness to archetypes, an ingenuousness of spirit—the freedom of the child.

CHAPTER VIII

I Can Only Tell You What You Already Know

"Good morning" is an everyday expression, made up of two words and eleven letters that provide little measure of its information content. Actually, the letters and words contain no information whatever unless they have both warp and weft. "Good morning" is a vocalization from a vast and complex world, without which these phonemes would be meaningless and devoid of any information. They presuppose an alphabet in which they can be articulated, then a vocabulary, a language and a syntax. "Good" emerges from a moral world and a convention of well-wishing, and "morning" from a star-governed time context containing within itself a cyclic vision of reality and much more besides. All this is implied and vaguely made to re-emerge from that two-word utterance, the sense of which depends less on the words than on the accompanying smile, or the slight closing of the eyelids, or the hint of a bow—all of which could suffice without the vocal accompaniment. Indeed, all of the above—both the binomial formula and the accompanying gesture—could be replaced by a raising of one's hat.

In other words, expressions are meaningless if they are not immersed in a context; rather, they express the context more than they express themselves. "Buon giorno!" means I am an Italian, I know you, I see you, I greet you; we're still on speaking terms; over and out. "Who's speaking?" Mallarmé would answer: "The word itself."

If we apply this sort of reasoning to what has been called genetic information—UGGCGUUCG, etc.—the same considerations hold. This gene segment (more correctly, this segment of mRNA) "means" tryptophan, arginine, serine, i.e. three amino acids. For tryptophan, however, UGG represents absolutely nothing. To bring them into relationship a strange cloverleaf molecule (tRNA) and a protein activator are needed, these being responsible for that mindless duty from the dawn of life, and in bondage thereto because it is impossible for them to evade this duty without causing the collapse of law and order in genetics. These relationships are 64 in number (each with a cloverleaf and an activator), and together they constitute what is referred to as the "genetic code." From a chemical standpoint the code is entirely arbitrary, a convention rather like that associating the symbols of the Morse code with the letters of the alphabet. No one except a telegraph key operator could guess that dash-dash-dot represents "g," just as no one could know without being told that UGG stands for tryptophan.

The arbitrary nature of the genetic code imparts to the whole of life a general character of mystery and indetermination. Molecular life is not a somehow foreseeable construction; it is not a machine with wheels and cogs but a system of elusive conventions, a contract that starts to mean something only after both parties have signed. Displaying the table of the genetic code as though it demonstrated the unveiling of life's interlocking puzzle is a mistaken enterprise, because life imposed its own constraints on itself without having or giving an explanation for doing so.

"Nature's conventions," i.e. rules not dictated by situational needs, are found at all levels of life. Just as in the genetic code and in human conversation, these rules are optional at the start and obligatory once they are agreed upon. They establish marks of identity and signs of recognition—among molecules, among cells, among individuals of a species or between different species.

A migrating population of butterflies flutters about, with individuals now associating, now dispersing, executing dance figures and pursuing courtships. Such a population must possess a system of recognition of group identity. A butterfly does not know who it is but does know to what group it belongs. Let us suppose it has brown wings with black markings. It recognizes as its companions all butterflies dressed in the same livery. The patterning may be devoid of any mimetic purpose, and may not serve to frighten off predators or absorb the warmth of the sun. It may have no "use" whatever beyond recognition of belonging. If the wings in the species were of a different color and pattern, no practical purpose would thereby be thwarted. The butterflies would still recognize one another, just as do butterflies of other colors. So it is with the colors of a sports team. It is of no importance that the uniform should be one color or another; but once the teams begin to play wearing those shirts, no player can put on other colors or he would be sent off the field immediately—not because the new shirt would be less suitable for the game as such, but because it contravenes the rules of what a biologist would call group identity recognition. We should not forever be looking for utility behind every living sign, since the sign may be no more than a convention or custom. According to Nietzsche, the origin of a moral code has nothing moral about it. The origin of a respected custom has nothing "respectable" or "customary" about it.

The insect that finds its way into a red, sweetly perfumed flower calyx to pillage the nectar, dusting itself with pollen before returning to the light in quest of another calyx like the first, may have no practical purpose behind its specific choice. Another shape, another color, another perfume might equally well provide it with sustenance, but this insect has chosen that shape, that color, and that perfume from time immemorial. These are its world and it knows no other. With its faceted eyes and its olfactory antennae this tiny inhabitant perceives a world entirely different from what we assume it does—perhaps a vast blue universe, a scent of incense, a music to which we are deaf but which harmonizes with the festive hum of its wings. That universe is known only to the insect. Living there, it has "generated a world," to quote Humberto Maturana, a world inhabited by it alone. The spaces and scenarios of this world are "conventional" because they are not the obligatory result of a choice made out of necessity. But neither are they the result of an agreement between a context and an intruder, because that universe did not exist before the tiny vagabond generated it and became part of it. The insect cannot abandon its universe without losing itself, because just as it generated its landscape, so the landscape generated the insect. "The mind and the world arise together," wrote Francisco Varela.

The identity of a species immersed in its environment requires an agreement between the living being and its habitat; but how can an identity be generated between a being without identity and an environment with no inhabitants? The "agreement" must exist before the parties; it is the agreement that creates the actors, the scene, the understanding. "Obedience precedes law," wrote Hector Bianciotti, "and in a sense, creates it."

Mayflies are like minuscule dragonflies that flutter about in swarms, millions of them, over watercourses. The adults have large eyes but no mouth; they have transparent wings and a slim abdomen terminating in long hoops. Linnaeus wrote that they have a short-lived joy, and often in the space of a single day celebrate marriage, birth and their obsequies. The larvae inhabit the water, where they swim or crawl about or anchor themselves to the bottom.

As soon as they wing aloft for the first time in flight, the newborn adults mate without a kiss, and the females swim upstream in swarms to lay their eggs in the headwaters of the torrent. The larvae emerging from the eggs let themselves be carried down by the current, and they spend a great deal of effort seeking the muddy environment that is to be theirs—the central, faster-flowing part of the stream, or the stagnant zones at the edge, or some area in between. Every species of mayfly settles in its own place as if on a checkerboard, guided thither by the instincts of the species, by an awareness of collective identity; and each one, in its place, pursues its own lifestyle—what Kinji Imanishi calls "sumiwake." There is no competition for territory, no differential reproduction. Imanishi speaks of a proto-identity, a sort of innate perception of identity that moves the members of every species to recognize themselves, to navigate toward the destined landfall and there deal with their short-lived existence and their sumiwake. Atuhiro Sibatani has likened this precise settling of mayflies to the movement of an embryo's cells in the direction of the various tissues that make up the organism.

Paleontology tells us that flowering plants and yucca moth insects came into existence at the same time. But paleontology concerns itself only with fossils, imprints and inclusions in amber. It knows nothing of the enchanted world that time has dissolved, and it will never know anything about it from analyzing traces of DNA.

Every living being carries within itself the capacity to generate a world, and it carries about with it the world that it has generated. What we call "information" is only a hint, or a complex of hints, to evoke that world. Different hints can ferret out one and the same world, and identical hints can lead to different worlds, because the world has a substantial autonomy vis-à-vis signs. Bull, taureau, rinde, Stier, toro, tjur, oks—all conjure up an image of the same animal. There is no point in racking our brains to decipher signs (whether in words or in DNA) in order to uncover the underlying substance of relationships among things. By the time we finish deciphering we will have arrived at nothing more than a dubious reconstruction of the history of signs. Between the text written in the DNA and the living form there is practically the same relationship as that between the words written in a book and the real world. The book is readable only because the world it describes already exists, ready to be diverted by some graphic stimulus into one or another watershed that has ever been shaping the landscape. All things have been said already, written already; and the sense of a book consists in reawakening dormant memories. A book tells you only what you know already.

"Before every human thought," wrote Jacques Lacan, "there would be a knowing, a system that we rediscover."

When grandma would retell a fairy tale by the fireside, the tale consisted not of sentences but of certain words, of sounds intoned by her voice, of the darkness in the room, of the flames and sparks flying up the chimney, of grandma's black dress, of the wind on the window panes, of the tick of the clock. And almost everything would continue to be what it had always been even if the words were to get lost when the listening child fell asleep. Children expect the tale already told, the loveliest and the newest. Yesterday's.

We were taught that a moonless night sky can express nothing, because the fixed stars are just placed there, without premeditation; it was Man who grouped them arbitrarily into constellations so that he could use them as reference points, or so he could attribute to them his legends—stories of heroes who did not exist before the time of myths and so could not be projected onto that supernatural vastness. Maybe what everyone knows here on earth went up there to put things in order, and the heavens are sending us back a secondary, already-known message. This is certainly what happened: It happened to me when someone—or a map—taught me to discern patterns in the stars. But supposing, instead, that I had met those stars already and was now learning, evoking them again? A supposition of this kind seems utterly useless and unverifiable. Would a blind baby be able to imagine and expect the heavens?

An experiment was conducted on birds—blackcaps, in this case. These are diurnal *Silviidae* that become nocturnal at migration time. When the moment for departure comes, they become agitated and must take off and fly in a south-south-westerly direction. In the experiment, individuals were raised in isolation from the time of hatching. In September or October the sky was revealed to them for the first time. Up there in splendid array were stars of Cassiopeia, of Lyra (with Vega) and Cygnus (with Deneb). The blackcaps became agitated and, without hesitation, set off flying south-south-west. If the stars became hidden, the blackcaps calmed down and lost their impatience to fly off in the direction characteristic of their species. The experiment was repeated in Spring, with the new season's stars, and the blackcaps left in the opposite direction—north-north-east! Were they therefore acquainted with the heavens when no one had taught them? "An animal with an innate image of the firmament!" commented Portmann.

The hypothesis defied all attempts to falsify it. When the blackcaps were placed in a planetarium, under an artificial sky, they repeated their astronomical performance. Another species of *Silviidae*, the lesser whitethroats, migrate toward the southeast, their species-specific direction, but south of the Mediterranean they abandon their south-easterly route and take a decidedly southern direction. If whitethroats flying in a planetarium are shown the night sky for the African coast between 15° and 10° N, the birds swing decidedly south.

On the vault of heaven at night the tiny *Silviidae* encounter the guide of the innate: They learn what they already know. Who can say how much of what we think we are learning is already there within us and awaits us from distant projections? But we moderns go for the ignoble solution—that all we possess was acquired or learned yesterday morning. Yet when I identified Ursa Major for the first time, didn't I already recognize it an instant beforehand?

Susumo Ohno is a Japanese biologist and supporter of the theory of evolution through gene duplication. Together with his pianist wife, he composed pieces of music "on the text of" certain genes that lent themselves to his purpose. He started by composing a code, an arbitrary code that translates nucleotides into notes, between which he introduced pauses and variations in volume, superimposing over the whole a highly imaginative accompaniment. In the hands of a clever pianist, the result was delightful. In this "composition," the DNA in the genes played a role happily analogous to what goes on in the cell. In order to express itself, DNA needs first of all a code for its deciphering. From this derive sounds that with a minimum of adaptation and enrichment express themselves in a musical world entirely foreign to the molecular world where they started. Ohno called this "DNA music" but clearly the entire world of music, with its instruments, its performers, its symbols, had to be there first; and DNA called for certain constraints that in the world of music could select sonorities acceptable as melodies. It is not that the piano and the violin were invented to perform music but that music adapted itself to the expressive potential of the instruments; or, rather, music and the instruments were born together. DNA is not a container of information whence music is born necessarily and automatically. Nor, by the same token, is it the primary container of genetic information.

Ohno also succeeded in carrying out the reverse operation. Starting with a Chopin nocturne he ran it through his code to reverse-transcribe a polynucleotide. In this way he was able to find (and who knows how many genes he must have examined) a gene segment which, once translated into musical notes, produced a melody very similar to the nocturne with which he started. Might not the genes have reassembled to reconstruct melodies that the world was already singing? Innate, ever there—it is music, and the instrumentalists learned to play it and, last of all, to transcribe it.

The behavior of a gene in trypanosomes illustrates how the cell, i.e. the world, makes up for the inadequacy of the information contained in its DNA. The enzyme formed by this particular gene (Cox) is the same in three species, but the gene in each is in one way or another incomplete and truncated. The tarantula's trypanosome gene lacks 29 letters (uridine), and that of the fascicolate trypanosome 32, whereas in trypanosome brucei the gene is reduced to a stub missing 60 percent of its letters. The cells of the three species repair the battered message after transcribing it into RNA, completing it by adding the lost letters (and removing one or two in excess) in such a way that a normal, active enzyme is formed. The cell thus knows the correct message and rectifies it as it proceeds if it ever occurs in the wrong form. According to these observations, genetic information is not like a ward where babies are born but rather like a registry office where citizens can check their vital statistics and make them complete again if any have been lost.

If my next-door neighbor, still half asleep, meets me early one morning and mumbles "… mornin'," I have little difficulty completing the truncated message, since the early hour and the encounter contain the message already.

This chapter calls for a postscript. If "I can only tell you what you already know," what point is there in my telling it to you? So many things that you know, that I know, abide in some limbo of forgetfulness or indifference, where we don't normally come across them. Discourse evokes a latent area of thought, a space where we can try to live, where it is possible for us to encounter each other, if only for a brief stretch of the way.

CHAPTER IX

PRESCRIBED
FORMS OF LIFE

⎯⎯⎯⎯⎯•⎯⎯⎯⎯⎯

The elegant spiral shell, the yellow corona of a daisy's petals, the gossamer spider's web—all are geometrically designed figures which seem to be solutions of algebraic functions unfolding in space.

Who has not admired iridescent, fleeting soap bubbles, their trembling clusters, and their silent return to drops of water? Life is not present in these bubbles, but they imitate cells and tissues with their round surfaces and adjoining walls. Bubbles are expressions of formulas and laws manifested in soapy water. And who has not been captivated by the design of snow crystals? These are gracious, tiny six-pointed stars—all of them hexagonal but all different from one another. I have always been particularly fascinated by the minuscule splash that a drop generates when it falls onto the surface of milk. This splash, which I saw for the first time in D'Arcy W. Thompson's *On Growth and Form*, is like an earl's coronet, with a band containing a score of tiny, sessile drops. I keep a copy of the photo on my desk, and sometimes take it with me to my lectures as the only illustration that I use. It is my modest heraldic crown—from the dairy.

What fascinates me so much about such a splash, something that is merely a split second in the cycle of turbulences caused by the impact of a drop on the surface of a liquid? In the first place, the drop is not the real "cause" of the emerging shape. A bead, or something else, will do instead; and it is too simple compared with the elegant sculpture it conjures forth. That shape, which reminds one of a sea anemone, contains no DNA and no genes, no memorized instructions. My little coronet is born of an impact, without any rehearsals or trial runs or natural selections. And then, perhaps most of all, I am moved by the fact that this coronet knows no time, save for that brief instant of its metamorphosis. For it has been the same since the time it first formed, since the time when, before milk existed, it took shape in dripping resin or mud. Could it not be described as a case of "autoevolution?"

The edge of a wave about to break seems to be animated by some frenzy to generate shapes. It becomes sinuous and unstable and crashes in a series of small jets that disintegrate in a phantasmagoria of mist-generating droplets. Then everything turns smooth again until another wave arrives to repeat the joyful turbulence. The powerful breakers that crash onto our beaches when a sea wind blows pile up on one another, one moment seeming becalmed and the next moment resuming their fury. In general structure each breaker has always been the same as every other, in this or any other sequence of waves, for as long as sea has been sea. Why do they have all the same shape? They are not relatives, they do not communicate amongst themselves, they do not reason; they just go berserk. They share a family resemblance because they obey the same law, a law brought by the wind that puffs them up and shatters them; but they are strangers to the wind, whose frenzy is greater than theirs. In other places the same wind produces whirlwinds and tornadoes, or undulations that move over fields of wheat without turning into breakers.

Shapes similar to bubbles or wave fronts are found in living things. Cells are like bubbles or bunches of bubbles. Honeycomb cells, too, are like spheres crushed up against one another to form hexagonal building blocks. An undulating, crenated structure—a cup with a wavy brim—is a common feature of many protozoans. The *Vorticellae* are sessile protozoans, formed on a stalk. They open upwards into a bell shape and develop crenations around the edge, forming a ring-like brim of lashes.

If the undulating patterns of *Vorticellae*, waves and splashes obey the same physical laws, this is not so important for our argument. D'Arcy Thompson's critics objected that he confused membrane tension with surface tension. What I want to stress here is that elaborate structures expressed by living organisms are arabesques that are extraneous to the biological complexity of the organism in question. A *Vorticella* contains more or less the same amount of genes and DNA as a butterfly or a cat. Genes and DNA serve to construct and reproduce the matter of the cell, but its form is little more than an open bubble that has taken in vital plasma. It could have made itself by itself. It is like a wineglass, and a wineglass is not made by the wine. The simplest forms of life and their living protoplasm are strangers to each other.

The shells of radiolarians and marine protozoa with siliceous skeletons come in a great variety of shapes, by which they can be classified into hundreds of species. Some of these have changed in no way whatsoever since the Cambrian—i.e., about half a billion years ago. Radiolarian skeletons are minuscule boxes. From the edges emerge tiny feather-like streamers, the purpose of which it is very difficult to fathom. Some have a polyhedric shape and reproduce the geometry of the platonic solids: tetrahedron, octahedron, icosahedron and dodecahedron (the cube, or hexahedron, is to be found in the skeleton of sponges). The sides of radiolarians have shapes like regular polygons, if somewhat curved; and on the edges one finds rigid verticillate spicules. What is interesting is that these solid forms do not appear in crystals, which are less regular. They are thus expressions of something living—shapes laid down in a graceful geometry forming a box around the cells, able to pass on their treasure chests intact to their offspring for millions and millions of generations. Other radiolarian skeletons are little cages shaped like hats, crowns, lobster pots and chandeliers. It is at the most microscopic and most archaic level of animal life that we come face to face with this plethora of vain embellishments and labored elegance. Life exhibits its morphogenetic imagination when nothing lays claim to it and no one is able to appreciate it, and preserves this for half a billion years in sleepy submarine deposits spread over vast expanses throughout the tropics.

Crystals or "flowers" of snow are in many ways similar to radiolarians. Under the microscope they look like six-pointed starlets with the most varied and gracious configurations based on the rule of six. They are like six halberds joined together at the base, or hexagons with their diagonals in relief, or six-petaled flowers, or six rods protruding from a central hexagon. They are different from the skeletons of the radiolarians because they have infinite variants that defy classification, because they do not reproduce and because they are not living. The radiolarians are more refined glassmakers; they can put together Murano chandeliers and crowns for Chinese emperors. One could say that radiolarians have added mass-production techniques and a touch of fancy to what goes on in the glassmaker's workshop. But the stark fact is still this: Mineral nature was an expert in shapes long before the world knew what life was.

Even the crystalline shapes of minerals begin with "seeds" or "crystallization nuclei." This is their art—to build up from minuscule beginnings such splendid glassware. A piece of crystal can shatter into seeds that nucleate fresh crystals from the substance in solution. Sometimes even fragments of substances different from the one that crystallizes can act as seeds. Alcohol can nucleate ice crystals. Chemists can go mad when faced with solutions of substances that have no intention of crystallizing: No sooner has the first crystal formed than the crystalline phantasmagoria explodes. Some people seem to have a special gift for promoting crystallization. A certain W. H. Perkins was famous for this, and people suspected that his unusual ability depended on his bushy beard in which—who knows?—invisible germs of crystals became lodged.

There are well known cases in which crystallization nuclei are organic substances. Bacteria of the genera *Pseudomonas*, *Enimia*, and *Xanthomonas* produce proteins that fold with a hexagonal symmetry similar to the nuclei of ice crystals. The presence of these proteins on leaves is responsible for the formation of frost. The frost "gene" has been isolated and modified so that it will no longer produce hexagonal proteins; it was then reintroduced into the host bacteria. When scattered over leaves, these altered bacteria hinder the formation of frost. Frost crystals or snow crystals are like ice shells crowning living structures.

The cell of a freshwater protozoan, *Stephanoeca diplocostata,* is capable of forming some two hundred bow-shaped rods of silicon hydrate. The tiny single-cell animal devours the rods and rapidly weaves them into a tiny silicon basket shaped like a lobster pot. It then discharges the basket into the water and uses it to capture the bacteria that nourish it. In order to crystallize the silicon, the protozoan secretes the right sort of crystallization nuclei from proteins, lipids or sugars—a microscopic basket-weaver, at work on the threshold between life and the mineral factory where crystals are made.

Precious mathematical edifices of high quality are produced by a group of very ancient protozoa, the *Foraminifera.* These microscopic beings form dainty calcareous shells, from the openings of which pseudopods emerge. Growth starts from a little chamber, which is almost sphere-shaped in the *Globigerina.* Beside this chamber the protozoan builds another larger chamber into which it extends itself. Then it forms a third and a fourth chamber, and so on, each larger in size but in the same position vis-à-vis the previous one. In this way a delicate shell forms, a volute in which the innermost loggia is the smallest and the oldest. The shape of the shell corresponds to an equiangular (or logarithmic)—i.e., to a cone that grows from the base and becomes wider as it winds around itself.

Gastropod mollusks are masters at building spiral shells. A closed curve is rotated around a fixed axis, remaining at all times geometrically similar to itself and enlarging its dimensions in geometric progression; it draws an equiangular spiral in space. With these slow and wise mathematical revolutions the snail generates its spiral shell, which it carries around with itself to the delight of the sea, the land and the malacologists. The nautilus, a cephalopod dating from the Cambrian, makes a shell that is divided into loggias like the shell of the minuscule *Globi-*

gerina. The last of the loggias, the biggest, is occupied by the animal it-self. That shell is the glory of living mathematics, designed as it is with admirable regularity and precision. It is found in the most ancient fossil layers, still intent on tracing its curves after half a billion years in the last species remaining as a living fossil.

The geometric structures we have been describing are not really living substances but mineral productions generated by animal cells. They do not grow; they are segregated and deposited as shields or breastplates, staircases or edifices, which the living body generates and makes its home. Likewise, the spiraling horns of the ibex and of many ruminants are external and non-living productions.

Living bodies with delicate geometric patterns are to be found among lower forms of life. Soft, throbbing shapes make a show of their grace and symmetry whether they are suspended in the water or lodged in sea or lake bottoms. The colored *Actiniae* stretch out their tentacles like vague, embalmed fountains; the polyps look like enchanted splashes; and certain medusas remind us of vortices, or pulsating bells, which when they "ring" liberate tiny eddies in the water. The corals produce rough-branched skeletons that form silent barriers, and from their red mass minuscule polyps dart out suddenly, like delicate white flowers. Starfish and sea urchins in their five-fold pattern grace the sea bottoms or the banks of seaweed with their astral symbols or their pincushions.

A jewel imprisoned between the shells of a mussel, "the sarcophagus of a worm" or of a grain of sand, the iridescent submarine princess un-veiled through the slit of a door, is the pearl that we leave to tumble out last—a symbol and seal of the vague, aimless and timeless beauty of the animal cosmos.

The entire plant kingdom, even in its highest forms, is a hymn to geometry. From the concentric rings of tree trunks to the regular shapes of leaves, to the spirals of phyllotaxis or the inflorescence of the sun-flower, to the sumptuous hexagonal or pentagonal symmetry of flower calyxes—all these things can be expressed as the development of simple algebraic formulas.

The leaf of the violet has a kidney-shaped profile that can be described by $r = \sin \theta/2$. A Grandi curve, based on the equation $r = \sin 5\theta/3$, describes the five petals of a simple flower, and a compound sinusoid of the type $r = (\sin \theta/2) - (\sin n\theta)$ draws the profile of a horse chestnut leaf. D'Arcy Thompson repeats the aphorism of Galileo (and of ancients such as Plato, Pythagoras, and perhaps Egyptian sages) that "the book of nature is written in geometrical symbols."

This brings us to the question to which this chapter has been leading us. Where does one look for those symbols and those formulas? Are they to be found in DNA, or in the dihedral angles of crystals, or in the surface tension of liquids? One answer could be that they are not to be found in any of these elements or in any other place, but are the abstractions of mathematicians—extracted from the confusion of the real world, if not invented afresh in their imaginations. This was not the thinking of Galileo or of D'Arcy Thompson, however. They attributed to these symbols and formulas a generative capacity, an efficacious activity. According to these authors "laws" are not in the world; they are perhaps modalities inherent in space that nevertheless exercise their activity in the world at their appointed places. They are unrealities that impose their rules on reality. I would simply add that those same laws or forms are to be found in our minds, too, enabling us to recognize them and to understand some aspect or another of reality. I would even go so far as to say that those laws and forms are the mind of the universe. (See Chapter XI below, "There's no body unless there's a mind.")

In other words, these laws hark back to a debate in the 1600s and 1700s over generation (dissemination theory), over whether the air—like a sunray holding its dust—contains the "germs" of all things flying about in expectation of alighting somewhere to form little embryos, or—rather—flowers, bubbles, splashes, crystals and pearls.

The central conclusion from all this is that in the emergence of forms "chance" plays no part, except perhaps to establish where these forms are going to turn up and display themselves. The forms that we have been describing come into being all of a sudden at the first attempt, and thus experience no "cumulative selection."

What is most fascinating in this story is the ecumenicity of forms and their indifference to variations of scale. The same spiral form can be observed in a galaxy, in a little eddy, and in the alpha helix of a protein molecule. Each of these is one-billionth the size of the one before it, yet all three are mathematical sisters.

Aristotle's three kingdoms—mineral, vegetable, animal—exhibit similar forms, which Antonio Lima-de-Faria, a Portuguese cytologist living in Sweden, sees as "homologous" and illustrates abundantly in his *Evolution without Selection* (1988). Foliations of gold and native bismuth have a structure with lateral veins at an acute angle, just like leaves or the underside of a wing of the *Kallima* butterfly (leaf insect). Dendritic structures are seen in native copper, in the RNA molecules deriving from DNA, in the branching body of the alga *Fucus,* in the colony of the hydroid *Aglaophenia,* and in the feathers of a bird. Frozen water vapor forms curves similar to the buds of ferns, to *Acanthus* leaves, and to the tentacles of a marine "flower" (crinoid).

Crystals of anhydrous calcium sulfate, colonies of a marine rotifer, and the floating body of the ascidian *Octacnemus* (a chordate) look like flowers. Chlorite crystals, mollusk shells, and the horns of a ram are elegant spirals that have nothing to learn from one another.

Lima-de-Faria concludes from all this that the universe is one great crystal factory and that shapes result from the organizing abilities of crystals and semi-crystals. Life is not necessary for shape, or shape for life. DNA, genes and chromosomes came later, when all morphological models were already present on the Earth. All these molecules did was to establish which variants of shape to fix and repeat; at most they have introduced into mineral patterns their biological products in order to touch up and stabilize the shapes. The basic shape of shells and mollusks is something for calcium carbonate to decide. The animal intercalates its keratin molecules in order to make the shell, as the case may be, into a long spiral or into a little globe.

Forms have a considerable autonomy residing in their generative nuclei, and life occupies and perpetuates the shapes already prescribed by the mineral world.

CHAPTER X

THE BIG DIFFERENCES ARE NOT DUE TO GENES

Glaucopian Athena, she of the blue eyes, carried a mutation in her genes for iris pigment. She differed from the rest of the gods, and from dark-eyed mortals such as Paris, because of a substitution in one DNA nucleotide. That she was a celestial goddess had nothing to do with her DNA. Differences in aspect or function regularly correspond to variations in DNA or genes, not in the stars.

That white forelock that certain gentlemen have over their forehead is a simple DNA affair.

All the characters of our physiognomy—height, long or turned-up nose, complexion, freckles, eyesight—are modulated by one or just a few genes.

Other, vaguer characters are modulated by a concerted assembly of genes. Most observed genetic modifications are indicative of disease or deficiencies with varying degrees of severity. Some of these, such as fair hair or Athena's eyes, can impart a certain grace to the individual, but most produce birth defects, a sad procession of disturbances and malformations. In the long alphabetical list of mutations in the mouse, one finds at the beginning "agitated," "alopecia," "anemia," "ataxia," ... and at the end "undulating," "vibrant," "waltzing," "whirling."

All these and a hundred other alterations have their basis in DNA; they are genetic misadventures that get perpetuated only because the biologist takes care to protect and preserve suffering for his research purposes. One spur to research on mutations was the hope that an accumulation of these might lead to a new species. But this never happened, because these are not the kinds of innovations that are worth keeping if we wish to construct anything. Closely related species have the same pattern of mutations, and this, if anything, is a mark of unity rather than a source of diversity.

If not mutations, then what is it that produces the great differences that distinguish one species from another, when each seems to be confined to its own universe, from which no accumulation of steps can ever deliver it?

Our question has been put in somewhat infelicitous terms in that it postulates a mechanism whereby a given species will turn into another as a result of an accumulation of minor shifts. To the extent that fossils can tell us anything, related species or groups do not descend one from the other but appear on the scene at the same time, as a result of some mysterious explosion or radiation, starting from one or another form that had maintained a certain ancestral plasticity, or had regressed to it. Sister species and related groups have substantially identical DNA, which undergoes differential variations (if at all) once the species have gone their different ways.

Examples of highly divergent forms possessing one and the same DNA are so conspicuous and so numerous that the marvel is that they have attracted so little attention. As a symbol of morphological diversity emerging from genetic identity we can take the caterpillar and the butterfly. There is nothing in which one resembles the other: The caterpillar is torpid; it crawls; it is usually dull-colored; its mouth has a chewing apparatus; its body is monotonously segmented, with all those hooks for feet. What we call metamorphosis is not really a change in form. Once the pupa or chrysalis stage is reached, the caterpillar starts emptying itself; its organs dissolve, and its outer covering is shed. Only certain groups of cells, called imaginal disks, remain vital. From these develop all the structures of the adult (the "imago"): antennae, stylets, proboscis, eyes, articulated legs, wings, and the fluttering lightness that warrants calling the butterfly "psyche."

Caterpillar and butterfly are widely differing forms, the one not derived from the other but both from totipotent embryo cells, some of which the caterpillar retains in its body so that they will in due course destroy it and replace it with another. The process of substituting the butterfly for the caterpillar is stimulated by adenotropic and ecdysic hormones and is repressed by a neotenic hormone. But the effects of these hormones are the simplest and most non-specific imaginable; and it is certainly not they that build the marvelous design of the butterfly on the corpse of a caterpillar. DNA may lend itself to such diverse forms, but it is not the DNA that imposes the blueprint, nor is it the hormones that do the organizing. Instead, it is one or more morphological destinies, lying in wait somewhere until they can one day reveal themselves.

Botanists have an even better appreciation for these double lives that carry identical DNA. Every plant has its regular alternation of generations—the haplophyte and diplophyte. In several algae these are morphologically almost indistinguishable, despite the fact that the diplophyte has twice the number of chromosomes as the haplophyte. In higher plants the diplophyte is represented by the adult plant, which inside special floral organs produces microscopic formations—the embryo sacs of the female and the pollen grains of the male. The embryo sacs nest in the bottoms of pistils, while the pollen grains, having left their anthers, fly about here and there like delicate dust in the air until they alight on the pistil stigma. From there they put out a slender tube-like structure that penetrates right into the embryo sac and fertilizes it.

Haploid nuclei from the mother cell (the oosphere) and the father cell (the generative nucleus) fuse together, forming a diploid nucleus that is the origin of the embryo. The embryo then remains enclosed within the seed until it bursts to produce the plant.

An embryo sac and a grain of pollen are full-fledged organisms, even though they have no other function than to unite in a micro-embrace. They are organisms by right, no less than the plant itself, even though the plant is millions of times larger than they, puts out deep-probing roots, and becomes a tall tree in the forest with a canopy of green leaves and variegated floral calyxes. The tobacco plant has 48 chromosomes in its cells, while its haploid pollen grains have 24. But this has nothing to do with the most majestic of plants or the tiniest of powdery lovers. No complicated expedients are needed to obtain plants that have 24 chromosomes but are almost indistinguishable from normal plants, or pollen grains with 48 chromosomes that are still like tiny, impalpable specks of dust.

These greatest of differences are unrelated to DNA, whereas the difference between a red-flowered plant and a white-flowered one is written with all fidelity in the DNA.

In many species the difference between male and female is a matter of chromosomes, yet the two sexes may be scarcely distinguishable. In some species, however, male and female have the same chromosomes (the same DNA) yet they are spectacularly different. In the marine worm *Bonellia viridis*, the female with its proboscis is one inch in length, but the male is almost invisible. The latter develops in the shape of a tiny crumb on the proboscis of its mate; the mate then swallows it, and the male lodges in the female's viscera like a minuscule sac, whose sole function is to impregnate its enormous spouse. The fertilized eggs resulting from this somewhat awkward copulation are all of equal size and wander around in the water. If they sink to the bottom they develop into a female, but if they land instead on the proboscis of a female they become that atrophied creature that is the male—which can scarcely be expected to exult with a "Vive la différence!"

Another case of marked diversity in beings with the same DNA is found in the caste system of social insects. In termites we see insects almost entirely lacking in any defensive system, what with their soft body, incomplete metamorphosis and no sting. Rather like humans. They have overcome these deficiencies with an organization that makes them some of the most fearsome inhabitants of the Earth. They build themselves colossal cities, impregnable termite castles complete with temperature and humidity control. These insects are divided into castes, the highest of which consists of a king and queen termite and a regiment of fully sexed individuals. In the hierarchy of citizens, these are the only perfect insects, with their faceted eyes and transparent, iridescent wings. They lead a life of leisure until their nuptial flight, after which they fall to earth in their copulating position, lose their wings, sink into the soil, and even lose their sight. A fortunate couple can go on to form another colony, where the king termite continues to lead his leisurely existence and fertilize the queen—who becomes enormous, flabby, and bloated with millions of eggs. The difference between the king and queen has its origins in the genetics of sex; but the difference between the royal couple and the worker and soldier termites—the common citizens of the colony—is not a matter of sex or DNA. Something else makes the worker and soldier castes very different from each other and from the reigning pair. The workers are whitish in color, dull-witted and blind, just like undeveloped larvae; and they work and work without interruption. The soldiers are likewise dull and blind, but they have what looks like an enormous helmet; and the head in various species is equipped with huge jaws, or a grotesque nose that can give off a resin-like substance to trap prey. The variety of forms and functions among the castes of termites are not due to any genetic differences; they all develop from the shapeless, uniform mass of eggs floating in the immense abdomen of the queen.

The startling differences between castes are in large measure due to arrested development. Soldier and worker termites are not the complete individuals that the sexed insects are. The former develop no eyes and no

wings, and they cannot look after themselves. To their inferior development they owe their willingness to serve and their meek discipline. They
remain part of the great being that is their colony; their only thought
is of the queen; and if the queen dies or is killed they remain paralyzed
where they stand, even if they are some distance away.

All this is sufficient to illustrate that the same DNA can be the origin of widely differing forms, and that the agent of diversification must
lie elsewhere. The agent may be a jelly, a hormone, or an ion, and it may
cause the arrest or continuation of development or alter some point of
bifurcation. Massive, far-reaching upheavals in the forms and functions
of organisms can be produced without bothering their DNA. The elixir
of variation must be sought elsewhere; and there is something magical
about it.

A classic example of major variations with identical DNA is found
in a diminutive Mexican amphibian, the axolotl. This animal has a tiny
white body and conspicuous red gills on both sides of the head. For the
local people it is a delicacy. In the early 1800s a specimen was put in the
lake in the Jardin des Plantes in Paris. At a certain point it changed into
a yellow and black salamander, lacking gills in the adult, the *Amblystoma
mexicanum*. The axolotl was the larva of the *Amblystoma*. In Mexico it
was held back at the larval stage, but it was able—and this is important—to reproduce as a larva and thus to form a species all on its own.
How did it turn into a salamander in that Paris lake? It was eventually
determined that the phenomenon was due to a simple fact: The Paris
water was richer in iodine, allowing full development of the hormone
necessary for metamorphosis. Axolotl and *Amblystoma*, as we have said,
have the same DNA; but the former has genes lying in a state of dor-
mancy (present for how many millennia, we may wonder) that enter into
action to produce the latter. The sleeping beauty with the crimson ruff
has awakened into a monster. The axolotl and the salamander had been
classified in different suborders of amphibians, but the difference between them actually lies in an insignificant iodine atom.

To complete our theorem that it is not DNA that makes the major differences, let us illustrate a few reciprocal situations in which the DNA differs yet appearances are similar. The cases in question are known—improperly—as "convergent evolution," in which different DNA is found in species that are so similar it is difficult to distinguish between them.

The marsupials constitute an order of their own within the class *Mammalia*. They made their appearance in the Cretaceous, before the true mammals (placentals) emerged in the Cenozoic. Marsupials come in a great variety of forms and life styles, which preceded and anticipated forms later expressed among the true mammals, as though they were a limited "first edition" of a subsequent large print run in the shape of modern warm-blooded animals (*Eutheria*). In the Australian forests one can still find a flying squirrel with a long tail and a webbed process between its fore- and hind-limbs. It is almost identical to the American flying squirrel, but the first is a marsupial and the second a true mammal. *Viverrines,* cats and dogs are revised editions of marsupials, and marsupial moles are a perfect anticipation of the moles we know today. It is not true that evolution never repeats itself; nor is it true that if evolution were to begin all over again, starting from some remote ancestor, it would follow ways totally different from the past. Just as in embryology, so in systematics there is a sort of law of "equifinality" in operation, according to which "a particular point of morphological arrival can be reached starting from different points of origin and following different development trajectories." (Hans Driesch)

A gerbil hops about on long hind legs, its small forelimbs clasped to its chest, as it disappears into the grass. I defy anyone to determine by sight whether it was a marsupial or a placental mammal. We cannot get around the difficulty by invoking "convergent evolution" that results from a natural selection of identical mutations. If this were so, we should be confronted with an improbable convergence in DNA sequences—something that no one has yet observed. Nature unveils its models, its potentialities—its "intentions" we might say—without any guarantee of a common substratum; and it reveals them in remote worlds, on distant pieces of DNA, by virtue of morphogenetic laws that favor certain morphological outcomes over others, almost as if these were the obligatory endings of fairy tales with different characters and different plots.

Japanese beetles branched out long, long ago into two groups ("clades") differing in the composition of their DNA. One group remained in the middle of the main island, while the other became distributed in outlying areas. Each group has given rise to three species, even though there has been no possibility of contact among them since the remote time when the two clades separated. The explanation? It could be that their common ancestor carried the three species within itself, unexpressed; and that since the exodus that separated the two groups its descendants have, on one front or the other, "learned" those innate forms and performed them in Nature's puppet show.

This way of seeing things has developed parallel to and independently of what may be called the Italian approach to evolution. In the early 1900s, Daniele Rosa of Turin postulated an evolution due to internal forces that work along the entire front of the species ("hologenesis"). The front could have split, and the transformation could have continued perpetuating itself with the same modalities along the various fragments of the front. Imagine children scattered among the different islands of a deserted archipelago. As the years pass they would be confronted with the same transformations in their isolated locations. They would come to know puberty, adolescence, love, thought, maturity and decline, all in the same sequence and with similar modalities. Maybe they would tell the same fairy tales. Daniele Rosa pictured generations proceeding for millennia toward their pre-established destiny, just as a child's future life unfolds before it. Such evolution along an entire front contrasts with the "diffusionist" vision of Darwin, in which innovation is strictly local and asserts itself in subsequent generations by reason of some superior capacity to reproduce, with the bearers of useful variations gradually replacing those less fit.

Darwinian diffusionism was unable to solve Hooker's paradox: How did similar species make their appearance in separate regions of the earth, beyond mountain or ocean barriers that prevented any communication among them? We are speaking of sister species that because of a lack of connections between their regions could not have spread via diffusion from a single center of origin.

Léon Croizat, a botanist of the Turin school who traveled throughout the world, advanced a theory to account for parallel evolutions. He called his theory "panbiogeography," according to which a species wanders at first over a vast area without undergoing variations. Then the area becomes divided by the emergence of insurmountable barriers. The segments of the species that have been separated by these barriers thereafter progress according to their innate transformation program and arrive at a similar morphological destiny unaware of each other.

Running birds appeared on different subcontinents in the Pliocene (about 5 million years ago) after the oceans had formed and the latter had separated as a result of continental drift in the Eocene (about 40 million years ago). Flightless birds cannot negotiate expanses of sea, so how could they end up running about the incommunicado plains of remote southern continents? Manqué birds, the *Struthionidae*, all have atrophied wings, a keel-less breastbone, bones without air sacs, enormous legs with a reduced number of digits, and a long S-shaped neck. In South America we find the nandù, in Africa the ostriches, in Australia the emu and in New Zealand the kiwi—all cousins in form, inhabiting far-removed continents to which they never emigrated. Similarity in form owes nothing to the environment or to genes but is the end result of a common trajectory that unfolds over time, like one and the same algebraic formula on so many blackboards of parallel universes.

It seems to me that the theory of hologenesis-panbiogeography, postulating as it does a historic destiny toward which species progress irrespective of their geography, has found a mathematical formulation in the convergence wave theory of Fantappié-Arcidiacono. According to this theory the shift toward order and complexity (syntropy) is not the result of local causes but of an overall design that summons reality to its destiny—an ostrich running in the outback of the future.

The same sort of complex structure has appeared many times in the long history of life, in organisms belonging to types far removed from each other, from different kinds of tissue and by differing processes, and underlain by DNA that is only distantly related. One need only cite the eye of the human and the eye of the octopus. Like the eyes in our heads, so the eyes of the octopus and the cuttlefish have a globe shape, a retina, a crystalline body and a transparent cornea. They stare at us from the side of their heads with a languid human look.

We may conclude that with the same DNA the most disparate forms (as in caterpillars and butterflies) can be made, while with disparate DNAs forms can be made that are almost the same (as in gerbils with a womb or a pouch), as well as equivalent organs (as with the eyes of the human and the octopus). Thus DNA relates to form only from a distance, and form possesses a fundamental autonomy vis-à-vis DNA. With a single information molecule all kinds of beings can be made, and with the most disparate kinds of information molecules the same end-result can be obtained.

Between DNA and an organism's appearance lurks the black box that conceals within itself the entire mystery of typologies and differences among organisms. The discipline that explores this *terra incognita* has been called "epigenetics." There is something infelicitous about the name, which still implies the theory that DNA, via a series of operations and interactions, when acting in a specific context, ends up producing an organism. Although it posits a distance between DNA and the final appearance of the organism, between genetics and form, between the genotype and phenotype, epigenetics is still about how DNA sets about "making" an organism. My view is that the problem is rather how the organism makes use of DNA, getting it to work or keeping it silent, and how it selects its areas of interest. DNA is not the starting point.

That said, we must be grateful to epigenetics, since it has resisted the tendency in molecular biology to collapse form onto DNA, phenotype onto genotype. Molecular biology, reaching its acme in DNA analysis, has always looked on the organism as a peripheral manifestation of DNA, as if the organism were merely a wrapping to hold what really represents life: the famous double helix, DNA. This new biology has had a predilection for making use of molds, bacteria, and viruses—microscopic entities possessing a minimum of form, in which defects in DNA (mutations) were immediately manifest in the cell's chemistry. Even higher organisms were demoted to become translations of DNA. Form was presented as something pleonastic, baroque, redundant, without which life could have done perfectly well—as it appeared to have done comfortably for three billion years, before plants and animals first made their appearance.

What happened when the marine and terrestrial paradise—which for billions of years had languished as a microbiological limbo—became populated with complex species and higher forms?

In a sense we could say that those genetic microbes—aerobic bacteria, spirochetes or blue algae—invaded forms and thereby established themselves as parasites or symbionts, producing by the strangest of symptomatologies that fatal malady that we call life.

Lynn Margulis formulated the Endosymbiotic Theory to explain the origin of higher (eukaryotic) cells. According to her theory, the cells of plants and animals are associations of different microorganisms that at one time were free. Cell organelles such as the respiratory mitochondria or photosynthetic plastids were once aerobic bacteria or blue algae, and the cilia and flagella were derived from spirochete-type bacilli. Even the elements involved in chromosome division (centrioles, kinetochores, and spindle fibers) are of microbial origin. When Margulis put forward her theory in the 1970s, it was considered a freak. Today it has earned general recognition, but it remains marginalized by Selection Theory because it dethrones Chance as the origin of the major phyla, and it emphasizes symbiotic collaboration over disputes governed by Natural Selection.

CHAPTER XI

There's No Body Unless There's a Mind

To utter the word "soul" in a discussion on evolution is to be guilty of the worst possible solecism. Once evolution entered into scientific themes there was no longer room for the soul. No room, because evolutionism attempts to explain things and their origin without invoking metaphysics; the soul, though it signifies a wind or breath, alights on us from transcendent lips. To exclude all discourse on the soul, however, is to tell the man in the street everything about evolution except what interests him. Well, then, the curiosity of the man in the street does not matter to the evolutionists. He will be invited to state his problem in different terms; otherwise he is kindly requested to occupy himself with other things.

But the problem cannot be made to go away like that. Every human being feels within himself or herself that there is a person, an "I," a light that each identifies with life and that each knows one has received into oneself at the moment of conception. Christians know that the complete person is born twice, once in the body and once in the soul. The most indulgent evolutionist may meet the Christian halfway, but he will insist that the soul, the thought, is a derivative or secretion of the body. "There can be no mind without a body." Thought and memory are but results of neurological connections.

The Catholic Church has given its assent, albeit cautiously, to the conventional idea of organic evolution. On one point alone has she taken a firm stand: Into the human body, at some stage in the hominization process, there descended the soul, from an otherworldly source. The idea that the soul is produced by the forces of matter, or is a mere epiphenomenon of matter, is something the Church cannot accept. Nor does she consider the idea dignifying. A papal message in October 1996 made it clear that at the origin of Man there was an ontological leap beyond the scope of physics to describe. "But the experience of metaphysical knowledge, of self-awareness and self-reflection, of moral conscience, freedom or, again, of aesthetic and religious experience"—these fall within the competence of philosophical analysis and reflection, while theology brings out their ultimate meaning according to the Creator's plans. Let scientists please confine themselves to the material basis.

The Church's position leaves a major biological problem wide open. If the Creator has placed the soul in the human body, at what point did He satisfy Himself that the body was sufficiently organized to receive a soul? And how had the organic evolution of Man proceeded up to the moment of this hallowing? How did a brain with the capacity for God form in a golem? How did the human body gradually achieve the likeness of Christ? The Church would not now be at a loss for answers if she had not gone along too hastily with the idea that, in the story of life, a hominid had completed the Darwinian course from an ape to Leonardo, that Man had gradually progressed from a knuckle-walking gait to an erect posture, to keep his backbone straight and his gaze upon the stars. My thesis all along has been that Man was born all of a sudden, in a great leap—i.e., in non-Darwinian fashion. His ontological leap was a biological leap as well.

How the leap happened, and from what species or primordial slime it started, science does not know—just as it does not know the answer in the case of any other fossil or living species. Indeed, if the transition were such that a newborn being became detached from a mysterious forebear, falling into an essentially new form, then the ancestral form is only of marginal interest to us and explains precious little. What interests us more is the shell within which the splash that generated our species organized itself. Man was born human, and he was not the offspring of a brute, certainly not by steps. As Heidegger put it, everything that is great is born great.

Having surmounted with this leap the obstacle of the bodily origin of the thinking being, we still have to deal with the relationship that what we call the soul established and still maintains with the body.

In the following pages I shall speak of "mind" rather than "soul," because if I can convey my ideas on the mind then only one final flight will remain—where most of my readers will not want to follow me and I risk burning my wings. We shall at any rate have escaped from the neurological swamp in which experimental science has painfully mired us.

The tiny embryonic ball begins to take form when the neural fold appears, burrowing its way from the rear in the direction of the future head and closing itself up in a neural tube. The neural tube then presides over the formation of various organs by causing the emerging nerves to branch out to them.

The neural tube is the motor center of the developing organism. It has an electrical activity that manifests itself in small, successive discharges. The embryo responds with little shocks and shivers as these discharges go about shaping the organs. This electrical activity generates the fibers and the circuitry that will eventually accommodate it. The nerve fibers transmit the impulses even before they are constituted; they are organized by the same impulse they carry. Organs begin to take their shapes and positions before the nerve endings are complete. They are sketched out close to the neural tube and are immediately joined by the nerves that establish their connections. They then shift in the direction of their proper locations and haul the nerves along with them, rather like underwater divers with their air hoses.

Once nerve fibers reach a muscle in the process of forming, they transmit impulses to the muscle that cause it to contract and relax. In doing work, the muscle takes shape. The reverse—a muscle waiting to take shape before it starts to work—never happens. With his highly refined experiments, Lev Beloussov demonstrated that morphogenesis is a process that depends on stresses and relaxations.

The energy-emanating nerve tube in the embryo can be likened to a mind that distributes its instructions to the organs and imparts to them those movements through which they form. The "ideas" disseminated by the neural axis have a precise plastic capacity, and they sculpt the tiny palpitating body according to an innate plan. If a nerve is damaged, local contractions stop and the no-longer-innervated organ ceases to form; and if the contractions are thwarted mechanically, the organ's development is halted. At a certain stage the toad embryo starts to fold up laterally in discrete clicks. When a thin glass probe is inserted from behind to impale the body, its movements stop and the intestine ceases to lengthen. (C. Bondi)

There is a stage in its development when the embryo begins to perceive stimuli from the external world. Its movements in response to these begin even before the sense receptors are complete. In this way the embryo begins its life of relationships. Messages from the outside world are vague and non-specific at first, and they serve only to give heart to an essentially autonomous morphogenetic program.

The human fetus begins registering the music that reaches it, and that it will be able to recognize after birth. Even the father's voice is "imprinted" and becomes familiar. The fetus begins to dream; and the "mind," no longer occupied exclusively with the self-organization of the body, readies himself or herself for receiving and elaborating models of thought. His or her morphogenetic "grammar" gradually becomes a grammar of the mind, a symbology.

René Thom, France's great mathematician and the originator of Catastrophe Theory, correlated the spatial differentiation of the embryo with adult functional activities. The embryo's morphogenetic gradients (head–tail, left–right, outside–inside) become gradients of activity in the adult. In the passage from the morphogenetic activity of the embryo to the functional activity of the adult, the so-called inductors or attractors that set in motion the morphological cascades are replaced by environmental evocators. Thus the inductors of the embryo's cephalic sensory make-up correspond to the "prey" that directs adult activity. The "mental" trajectory of the predator animal is already prepared in the blastula stage of the early embryo. Sensory, motor, digestive, and excretory organs prefigure the functions of the adult hunter: perception of the prey, pursuit, devouring, digestion, assimilation, and excretion. The minuscule embryo is preparing for a destiny and the morphogenetic processes anticipate the direction it will take in its life of relationships. Seen in this way, the phenomena of consciousness and ideation emerge as an external projection of prior morphological elaboration. Forms and ideas are ontologically linked. According to J. M. Lotman, at a certain level of abstraction the "thinking structure" and the "living structure" are defined in the same way. For V. I. Verdansky, intellectual activity is "a potent biological force," an emergent state of the biosphere—the "noosphere." "The noosphere is a new geological phenomenon on our planet."

In the early embryo forces are activated that evoke the form of the body, bringing it into relationship with the outside world and projecting it there to "construct a world." As formative activities, all these processes can be considered "mental," and their sequence can with good reason be called a "journey of the soul."

The soul is that excitable little something that awakens when sperm and egg embrace. It calls forth nascent forms, and by those same forms makes itself welcome. It listens to the outside and elaborates what it perceives in mental figures. So it dreams until the day when, emerging into the light, it sets about constructing a world, a world that is both the expansion of itself and its temptation. "Animula vagula, blandula." (Hadrian)

The lizard, too, has a little soul, a world of its own, but this does not generate ideas. At most, it will grow a claw or tail again if either is cut off. The planarian worm will even grow its head again.

Let no one think I am here to thumb a ride for that delicate journey of the soul along the highways of biology. If I decide to deal with the soul, I realize I am casting doubt on the very notions of material substance and time.

When the "I" becomes conscious of the soul, it also acquires a duty to carry the job through to completion, to continue the flight. The danger lies in the sin of refusing life to the soul, or song to the innate. Face to face with the word "soul"—we must accept completely the risks of the words we use—the "I" runs the risk of dissolving, along with its claim to singularity and locality.

To the assertion "there's no mind unless there's a body"—no thought unless there's a brain—I assert the contrary: "There can be no body unless there's a mind," no form unless there's a founding principle, no lizard's tail unless there's the idea of it. Buddha, we are told, expressed a similar idea when he said, "Everything possessing form exists thanks to the mind." For the Hindu religion, every object is the materialization of an idea. "In the beginning was the Word," St. John's Gospel begins. "Through Him all things were made.... And the Word became flesh."

This world of ours is really us, the projection of the force that molded us and offered us to the light, to consciousness, to memories. What autonomy, what permanence does it have? And what does such a world become after the death of the body? Will it remain there, incorporeal yet personal, a soul without a body, to remember the lost beloved, to accompany migrant swallows, to chase after fugitive galaxies until it ends by mingling itself with the spirit of the universe?

What use is it for a person to have known the solitary savannahs of life, the moonlit profundities of the night, the joy of love—if, in the very moment of his passing, he gets confused and his soul, afraid and uncared-for, takes refuge in some hopeless corner of a room? Why practice the ways of consciousness, why set oneself up as imago mundi, why engage in colloquy with eternity—if, at the end, all that awaits us is a shipwreck, forgetful of everything except suffering in the inhospitable sheets of an unmade bed, slumped on an uncomfortable pillow? Surely it would be better not to try moving beyond that ultimate tunnel, dragging with one the luggage of the years. Better to concentrate on the final brief encounter: a sparrow on the balcony rail, a flower in some antique fable, a picture on the wall, a smile.

If it is true, as shall I assert in the next chapter, that every part of an organism contains the whole, that every idea reflects the world, then the last step will prefigure the eternal.

CHAPTER XII

MATHEMATICS FOR TALKING ABOUT CLOUDS

The most graceful, elegant and gentle of all natural forms—flowers, seashells, crystals—can be sketched mathematically and reproduced by developing algebraic formulas. This suggests that the simple forms are derived spontaneously from generative principles, triggered by chemical or vital processes operating in the material. Calcium salts accumulate in the enlarging hem of the snail's mantle, with activity at a maximum on the circle's arc and declining regularly until it fades on the opposite arc. The result is a cone that is twisted and folded back on itself. A tree trunk grows from tiny channels arranged all around it under the bark. It forms a cylinder which, when sawed transversally, shows a regular series of concentric rings, with one ring for each year of the tree's life and a scar here and there from the buffetings of time.

The simple beauty of nature is obedient and a bit naive: It allows its secret to be understood and described, because it is the development of an exercise in elementary algebra. What sort of landscape might nature present to us, however, if it were just the daughter of a few rules; if mountains were cones, if coastlines were arcs of circles, if tree trunks were as smooth as columns, and if lightning flashed only in straight lines? (Mandelbrot) There would be little attraction to living in a landscape composed of cylinders, spheres, and pyramids à la Cézanne.

By the grace of Heaven, nature is erratic and elusive, undisciplined and unpredictable; and the god of geometry has allowed free rein to impertinent fauns and flighty nymphs. Unruliness is the rule in nature, but it is not a variant of chaos; and the most complex and bizarre forms conceal the elusive rules that govern them.

To subdue these hidden rules mathematically is a form of arrogance that Nature will not tolerate—a vain effort bound to disappoint. Nevertheless, an approach is open to us, a humble and playful one, whereby we can mimic nature's morphology. Using graphics, especially now that we have computers, we can develop one or a hundred or a thousand sketches, then we can look among them for the closest fit to nature. Of these we know the design, of nature we don't.

In the mid-1970s, a family of patterns called "fractals" (Mandelbrot) attracted much attention, for reasons perhaps difficult to understand unless you are a mathematician. Even though the patterns unfold on a page or a monitor, they do not have a whole number of dimensions, and a fraction of their surface is always left free. Another fascinating property of fractals is their "self-similarity," which is due to the fact that fractal figures are produced by the reiteration of a simple rule, such as breaking the central part of a segment with an angle and then breaking the middle of the resulting line with smaller, proportionally sized angles, and so on. What emerges is a complex figure in which every part is a miniature reproduction of the original figure.

One well-known fractal is the Peano-Mandelbrot "snowflake." It starts with two superimposed triangles pointing in opposite directions so as to form a six-pointed star around a hexagon. A black figure is drawn from one side of this hexagon, limited by a fragment; it invades half of one of the triangles and looks like the shadow of a grasping hand. On all sides of this "hand" the figure is repeated outwards, on a smaller scale; and, toward the inside, the same figure is repeated in white on black. Continuing in this way, the star develops into a minute marquetry of hands and smaller hands, black ones and white ones, producing the appearance of a snowflake with the following property: Each detail is the same as the whole and, if isolated and subjected to the same rules, is able to regenerate the whole in miniature.

Take a rectangle and draw on its shorter side smaller, oblique and adjacent rectangles. On the shorter sides of these draw smaller, proportional rectangles and continue doing so. The result will look like a minute fern leaf.

By taking simple rules of variation as the starting point and applying them repeatedly, one can watch the unfolding of fractal figures suggestive of mountain profiles, coastline patterns, Christmas trees, estuaries, or respiratory and circulatory networks. The landscape is no longer the surrealistic one of Euclidean geometry; instead it looks animated, earthy, uncalculated. If we introduce into the process small, accidental errors—scars left by history—anomalies emerge that bring the images closer to our own familiar experience. Imagine that you are travelling along a road lined on one side by factories and on the other by a forest. You would see on one side geometry, and on the other a fractal scene.

This whole family of figures consists simply of simulations; the figures do not necessarily develop according to the rules adopted by nature in building the originals. The figures tell us, however, that nature does not need complex rules or principles to construct bodies that are articulated and bizarre, or splendid and tortuous like Chinese dragons.

Once fractals began populating the computer world they produced no end of wonders—indescribable forms and colors, all derived from the monotonous and repetitive aggregation of elementary patterns. Why do we allow ourselves to become captivated by these deceptive figures, these fascinating simulations? They tell us that nature must have its rules of composition if it is clever enough to imitate its own imitations. They tell us that it is possible to paint the town red beginning from simple algorithms, with no need for megabytes of information encoded in the DNA of organisms.

Fractals, it has been said, are mathematics for talking about clouds. A cloud is an accumulation of clouds, each one of which is an accumulation of smaller clouds. And each small cloud, if we look carefully, is made up of so many wisps of clouds, especially at the margins. One such wisp, duly enlarged, is like the mother cloud. Clouds can be globular, fleecy, or stratified. They can fill the entire sky or they can drift like feathers in a serene blue vastness. Black, gray, white, orange—they can look like infernal mountain chains, smoky masses, paradisiacal savannahs, big fat cats, bucking broncos, or whipped cream. If we want to know the world around us we must contemplate clouds, because even if we forget to watch them it is the clouds that are queens of the landscape; a country is beautiful if the clouds above it are beautiful. They reveal their minute composition by dissolving into droplets or, again, by flaking into frozen water to form blankets of snow. According to Élemire Zolla, clouds are archetypes of reality.

Like clouds, every fractal body has structures that are similar at different levels of analysis—a property known as "invariance of scale" or "self-similarity." I would not have mentioned fractals if this property of theirs were not the essential quality of life. A plant is "self-similar." A twig, a bud, a seed, even a cell, can regenerate the entire plant; it contains the secret of the whole. Life is made up of countless, unexpressed designs for life. "Alles ist Blatt," said Goethe, meaning that the tree is like a leaf, and all the branches and twigs are like the ribs of the leaf, and all the petals, stamens or pistils are variations of the leaf. Everything is a leaf.

Let us honor with a glance the humble cauliflower. Its great, swollen, pale green cone, tucked in a leafy boll, is a juvenile inflorescence. It is made up of numerous small branches, each bearing a cluster of floral buds. The cone itself has a marvelous fractal structure, consisting of a hundred or so minor cones arranged along spiraling ribs, each of which is made up of other cones, likewise on spiral ribs, themselves consisting of microscopic cones.

Every detail of its sumptuous structure is a diminutive whole, a cauliflower in miniature. Each component cone is self-similar and invariant regardless of scale.

The fractal structure suggests an image of nature as omnipresent structure. Every corner of the world is the entire world, and the entire world is in each and every corner to a greater or lesser extent. The never ending problem of the whole and the parts finds a solution that is at once recent and ancient. In a blade of grass is the universe, and the universe—let us try to believe—is a blade of grass.

Let me just make a passing reference to what is called the reductionist or atomistic conception of nature, the view that any given thing can be understood by examining its component parts, by taking the doll to pieces. According to this view, an engine or a gun can be understood by taking it apart and studying the function of each piece. Similarly, a living thing can be understood by opening it up and examining its organs, its bodily machinery, its glands, and so on. The significance of the individual piece, in turn, is revealed by dissecting it down to the level of tissues, cells, molecules, and atoms. Once the molecules are understood, one should, in theory, be able to reassemble the whole organism. Obviously, this is materially impossible and no one even makes the attempt; but the central assumption of the reductionist way of seeing things is that, in the abstract, it is possible.

Opposing the reductionists are the holists (from the Greek *òlos*, whole; sometimes improperly spelled "wholists"), who, as their name implies, are interested in the whole. Their assumption is that "the whole is more than the sum of its parts." In other words, movement from the parts to higher levels of reality is not automatic. Molecules alone do not make a cell, and organs do not explain the organism. Emerging life demands something new, something not encountered before, something that its parts know not. A new event is needed at every step in the origin of cells, forms, animals, intelligence, society—an event we might call an invention of nature, a creative act, an evolutionary leap forward; in any case, an event neither contained nor foreseeable in the conditions that precede it, nor in the component parts. The glue that binds the parts together is something mysterious and arcane, and it is not produced by the pieces themselves. Holists go further and say that not only is the whole greater than the sum of its parts, but also that the parts are meaningless except as subspecies of the whole: Only by virtue of its place in the whole is the part what it is. The emergence of a higher level of complexity makes sense of the parts and endows them with a function, a value they previously did not have. The eyeball without a nerve and brain behind it is biological nonsense. Before the whole comes into being, the parts are meaningless objects, or they are possessed of a different meaning of their own; they are clearly not parts of a whole. They are "minor" wholes, wholes for their parts, wholes for their "sub-wholes." A more rigorous consideration would lead one to conclude that the parts do not precede the whole but are born and acquire meaning along with the whole. Their role as "parts" is a role in the cognitive process but not a role in the generative process.

Empedocles imagined a chaotic universe, in which heads, necks, hearts, and feet wandered about aimlessly in space and came together by chance to form all kinds of beings; but only those beings survived that had their parts in the right place. The philosopher gave the name "love" to the force that brought the parts together. But Empedocles' "love" has nothing to do with what the holists call "emergence." Outside the holistic whole there are no parts, or they have no meaning; parts follow the whole or emerge along with the whole.

François Jacob solved the problem of the parts-in-the-whole with a cunning sleight-of-hand. Nature, he says, works as a tinkerer, rummaging among warehouse discards or rubbish and patching together what it finds, assembling hitherto unseen kinds of apparatus which form novel structures that get added to the oddities already in the world. According to this view parts anticipate the whole, though they come from other wholes. They have been dumped along the roadsides of the past and subsequently restored—perhaps to their original function, or to some other function altogether. And so it might be that the same mechanisms are found in all sorts of different objects, just as the same biochemical pathways are found in the various faunas of our biological zoos or the floras of our botanical gardens. This is no reductionist image, because the parts do not "prescribe" the whole; nor is it holistic, because the parts are said to come first, sometimes changing their roles to become different "parts."

In the fractal perspective, all games are turned upside down. In a "self-similar" structure every part is the whole, containing within itself the ability to generate the whole. Each part is the whole on a smaller scale, the germ of the whole. Life is an association of germs, each ready to express much more than it actually shows.

Life is anything that has the capacity to regenerate itself from one of its parts. Life grows by self-reproduction. It scatters into the world the buds of which it is made, bestowing on the landscape scions of itself, and invading the non-I with its own self-reproductions.

True, there can be no reproducing a man from a hair, a woman from a curl. In higher beings, the ability to make self-similars, to regenerate, remains the prerogative of specialized cells—stem cells, and ultimately the ovum. But those properties are present, concealed and unexpressed, in every drop of life. Higher forms of life possess a sublime power: They die. The egg is light, and death is that shadow that takes light away from late and peripheral parts of the organism. By measuring out its darkness, death creates contrast, nuance, and form, and these define the body. Every oocyte contains within itself numerous potentialities (which conceal each other), and it knows how to construct a dying fractal—the body of a mother cat surrounded by her kittens, her self-similars, her eternity.

Lower organisms—algae, bacteria, and yeasts—belong to the monotonous tribe of the immortals. They progress in cycles, they split, they bud; but they always remain at the same level unless they stop separating and start aggregating, producing self-similar bodies. Like fractals.

And when the time comes for them to die and their development ceases—it is then that history begins, and forms originate. Then sin comes along to inaugurate thought and consciousness, which are the expression of life in ideas. This is the mind's attempt, as vain as it is illicit, to conceive of the world as its self-similar.

CHAPTER XIII

WHAT TEACHES PROTEINS THEIR SHAPES?

———————•———————

Late twentieth-century genetics has met with two failures, both of them a result of its reductionism and its predilection for artificiality. These are recorded in fields so far removed from each other and in organisms so remote that any connection between them has gone unnoticed. The first is the failure of genetic engineering; the second is the crisis in Britain's cattle industry as a result of the "mad cow" epidemic. In both cases, protein anomalies or diseases emerged that biological theory had not foreseen. What calls itself (not without a certain irony) the Central Dogma of molecular biology claims that DNA, the self-reproducing molecule of heredity, composes proteins. The latter have no self-reproducing ability. Proteins, the Central Dogma says, have to be continuously supplied to the cell in accordance with instructions contained exclusively in DNA—by the genes, therefore, since genes are made up of DNA. All heredity is in the DNA.

As its name implies, genetic engineering had the ambition to extract a gene (a length of DNA) from a higher organism (a human, for example) and transplant it into a bacterial cell by means of an appropriate carrier. The experiment succeeded in the case of genes for insulin and growth factor, and it became possible to make sufficient quantities of these to market. These fortunate beginnings, however, were followed by widespread disappointment. Proteins transferred from one kingdom to another through biological smuggling did, indeed, form in their new environment, but they often failed to function. The amino acids making up the proteins duly placed themselves in the proper sequence, but they failed to take up the spatial configuration necessary to make the proteins as active as they were in the donor species. Instead, they formed glutinous masses, the equivalent of a dish of overcooked spaghetti. After a promising start, biotechnology companies were formed all over the world (and quoted on stock exchanges) with the object of fabricating proteins old and new in brewery fermentation tanks, without recourse to the slaughterhouse, at negligible cost. But crisis soon overtook the system, and many companies went out of business. "Fueled by hopes and dreams," the *New Scientist* commented in July, 1998, "rather than actual products, [the biotechnology industry] was a house of cards waiting for a scandal to bring it crashing down." And that, of course, is what happened.

How was it that technology failed where nature had succeeded? Why was it that smuggled or imported genes did not form the proper proteins? The problem is that in order to get proteins to function one needs not only an orderly and correct sequence of amino acids, but also a spatial configuration that folds them into the proper association with each other and enables them to interact with the molecules on which they are supposed to work. Jacques Monod was specific on this point: The spatial information necessary for specifying the three-dimensional structure of a protein is vastly greater than the information contained in the sequence. Monod resolved the "paradox" in Darwinian terms by saying that the initial conditions provided by the sequence do not specify the correct spatial structure but merely eliminate some wrong ones. Within the cell there is no sign of a massive slaughter of mangled proteins. The proteins are perfectly well settled in their proper places, where they form coherent systems by working together closely and harmoniously. In order to arrive at its appointed place unscathed, the nascent protein thread must first solve its folding problem.

Toward the end of the 1980s, several molecules were discovered that help nascent proteins to defend themselves against bad influences in their subcellular environment. The molecules, which are proteins themselves, form a sort of envelope inside which the nascent protein moves toward the place where it will take on its definitive configuration. The boxes work simply as a container for the proteins and have no role in the latter's folding process, which commences once the protein has parted from its "chaperone," as this has come to be called. If there is no chaperone, whether due to a defect in the cell or because the nascent protein finds itself in a foreign cell environment, the jilted one takes refuge in questionable relations. Dozens of classes of chaperones have been discovered; each one can protect different proteins, but none of them make any contribution to the shape those proteins will eventually assume.

What sort of cellular structure promotes correct folding? What structure is informed about all the twists and turns that various proteins must go through in order to become active? Chaperones prevent proteins from taking wrong turns. But what teaches them the right ones?

An indication that a protein's spatial structure is determined by contact with a sister protein has emerged, unexpectedly, from research on "mad cow disease." In this affliction, contraband proteins penetrate the cow's brain cells, transmitted not by genes but directly by ingestion, since they are resistant to digestive enzymes. Their arrival causes an upheaval in the spatial structure of indigenous sister proteins. The latter do not change their sequence but rather their "attitude" in space, which they alter by conforming to the ingested proteins. The phenomenon amounts to a hitherto unexpected "heredity by contact"—rather in the same way that bad company makes a child behave badly (or a badly behaved child makes company bad), or in the same way one rotten apple spoils the others in a basket.

Protein heredity by contact was first postulated by Stanley Prusiner (Nobel Prize, 1997), and it was considered by many to be the most important event in biology in the half-century after the discovery of the structure of DNA and the genetic code. Prusiner developed his ideas while investigating the etiology of mad cow disease—a contagious disease that brought the United Kingdom's stock-raising sector to its knees. Cows with the disease lost their sense of direction, developed a stumbling gait, had difficulty remaining upright, and ultimately fell over dead. Their brains were found to be riddled with tiny cavities, so the disease was named bovine spongiform encephalitis (BSE). Attempts to find a causative virus or bacterium were unsuccessful, and Prusiner's hypothesis that the causative agent was a protein eventually gained widespread acceptance. It conflicted, however, with the Central Dogma that heredity is transmitted exclusively by nucleic acids. All efforts to find nucleic acids (DNA or RNA) in the infectious agent failed. Infectious proteins were inadmissible under the Central Dogma, and their autonomous transmission was out of the question. And yet.... And yet, the protein was eventually found and isolated. It was named a "prion," and its mode of action was discovered and reproduced in vitro.

A pathogenic prion has the rough form of a normal protein. The BSE prion has an anomalous fold known as a β-sheet where the healthy protein has an α-helix. The latter forms a myelin sheath around the nerve cell membrane. All normal proteins have the same orientation and winding pattern. When prions (β proteins) enter the cell, they insert themselves into the palisades of normal (α) proteins and imprint on them their distorted structure, which then spreads throughout the cell and from cell to cell. The infected sheaths form aggregates inside the cell and disorganize the coherent regimes of the cell membranes, giving rise to cavities in the brain and madness in the cow. What has happened is not the multiplication of a pathogen, and not a true case of molecular heredity, but the transmission of a deformity in the molecule arising out of contact with another deformed molecule—a "rotten apple" effect.

Prions responsible for the mad cow syndrome came from other diseased cows, via feed that contained protein powder obtained from slaughterhouse waste. At the root of the disease was a sort of cannibalism that had been forced on cows removed from their grazing and summer pastures. It is suspected that a rare brain malady in humans, Creutzfeld-Jacob Disease (CJD), may be traceable to meat from mad cows.

Some cases of protein heredity by contact have been reported in yeast. A certain protein (Ure 2) may take on an anomalous form in certain cells. The deformation spreads to healthy Ure 2 proteins and is transmitted to all bud cells and the entire progeny of crosses with normal cells.

The initiation of (tertiary) protein configuration by the organization of a pre-existing protein has been elegantly demonstrated in the flagella of *Salmonella* bacteria. These flagella consist of aggregated protein subunits (flagellin) that assume a spatial configuration in the form of wavy or spiral whips. Once flagellin is disaggregated and dissolved, it does not reorganize itself spontaneously. If a tiny fragment of wavy flagellum is added to the solution, however, the dispersed molecules organize to form wavy flagella. If instead a fragment of spiral flagellum is added, spiral flagella form. The source of the flagellin subunits is of no significance whatsoever.

The *Paramecium*, a protozoan of the infusoria, has a body covered with cilia that all point in the same direction. Tracy Sonneborn (1970) used microsurgery to invert a row of cilia so that they pointed in the wrong direction. When the protozoan reproduced—i.e., divided in two—the inverted cilia obstinately retained their contrary orientation even after renewal of the nucleus (DNA). Sonneborn proposed the hypothesis that underlying the cilia was a coherent structure that self-reproduced even in its inverted configuration.

Some development biologists have suggested that changing the configuration of certain proteins may play a role in embryonic differentiation by determining whether a particular group of cells is to become liver, muscle or some other tissue. In this case, the various organs would be analogous to "pathologies" transmitted to tissues by prions that modify protein regimes.

Research on protein continuity has thus revealed a parallel heredity that operates side by side with DNA and is no less rich than it. This revolutionizes the Central Dogma's concept of heredity because it disregards the letter sequence of proteins, but is concerned instead with their attitude in space.

Heredity by contact has other noteworthy qualities. First among these is that it can enter into transactions with the environment and even with diet. It can undergo modifications other than those due to chance, which are an illness of DNA. The second noteworthy quality is that it can be passed on, as an ailment or as an impression, not only from ancestors to descendants but also among neighbors and acquaintances.

In any case, the real problem facing us is the stability of forms. Prion-induced anomalies, by showing us the exception, have provided evidence of a fundamental stabilizing mechanism, a regime of structural coherence in sister proteins, and a morphological solidarity that extends to vast areas of developing tissues.

The animal organism is derived from the fertilized egg. The generative principles of protein systems, like crystal seeds, must pass through the egg. Genetics provides us with many examples of what has been called maternal or cytoplasmic heredity. Such heredity constitutes the most important exception to Mendelian inheritance, the latter making no distinction between the parent, male or female, from which the trait in question is derived. Where segregation among hybrids does not take place, and all progeny have a particular character from the mother, the cause might be cytoplasmic (protein) structures to which the sperm makes little contribution. The best known example is that of right- or left-handed rotation of the shell of the snail *Limnaea peregra*. The progeny always have shells that rotate with the same handedness as the mother, irrespective of the father. (A gene mutation has been described that interferes with the phenomenon and thus makes it possible to study it.) This remains an excellent example of spatial configuration that is entrusted to protein continuity—via the egg of the mother. The attitude of the researcher must not be one of hunting at all costs for a gene that alters a process, in order to bring it back into the "DNA is everything" fold. Protein heredity is vast and pervasive, and the fact that it is difficult to study argues for its ubiquity.

Genetics is the science of the mutable. The immutable is not amenable to its jurisdiction or its algebra. Genetics disregards the mother's essential and pervasive contribution, which is steady and unchanging. The "DNA is everything" dogma neglects maternal generosity and constancy.

CHAPTER XIV

LIFE IS A GRAND BALLET

In physics, the first thing we learn in school is that matter comes in one of three states: solid, liquid, gas. We know that water can be ice, or its liquid self, or water vapor, depending on the temperature. There is also a fourth state of matter, intermediate between solid and liquid, between the hard crystal and the elusive fluid: It's called "liquid crystal." This intermediate and changeable state is used in digital clocks and computer screens. Its iridescence and temperature-sensitive color are exploited in "instant" thermometers and even in the fashion world. As the temperature varies in different parts of the body, a garment's fabric will take on different colors and shades. The fashion model is doubtless unaware that the internal tissues of her body, when observed under appropriate lighting, are iridescent and changing just like the tight-fitting dress she is wearing.

Most biological structures are in the liquid crystal state, where they hold themselves solid yet not rigid, as befits life. All cell membranes are liquid crystals; so is the DNA in a chromosome. So are proteins in general, and the proteins of the cytoskeleton, muscles and connective tissues in particular. If we picture protein molecules as so many small rods or stakes, a solid protein crystal would turn out to be a long palisade, extending over three dimensions, in which each stake has its proper position and all stakes point in the same direction. If we raise the temperature the bonds loosen and the positional order is lost, but the molecules remain tenderly joined and loosely parallel, still pointing in the same direction. Such a flexible, coherent palisade is called the "nematic phase" of the liquid crystal. Raise the temperature further, and all semblance of order vanishes; the molecules, now directionless, wander about freely and form a chaotic fluid. Liquid crystals are ordered, yet they are mobile, flexible and reactive. Every molecule imparts its orientation to its neighbors, and the uniformity of the ordering can extend to millions and millions of molecules. Order is not decided by the composition of the individual molecule, even though all molecules in the same assemblage have the same composition. Molecules learn their order from the company they keep, and the company receives its order from distant molecules—in much the same way that a community adopts norms that were laid down when and by whom nobody knows.

Some molecules (such as cholesteryl benzoate) look like screws, which may be right- or left-handed. Here again the molecules in a liquid crystal show their solidarity: A new arrival takes on the handedness of its neighbors, joining a palisade of screws, as it were—all of them turning in the same direction, like so many parallel spiral staircases.

Liquid crystals have a three-dimensional structure unknown to the nascent protein molecule, which learns about it from the company it joins at the end of its route through the cell. This is the school where the molecule takes on and conserves the folding pattern that will bring it to life and make it functional.

The membranes of nerve cells on which prions wreak their havoc are liquid crystals into which deformed proteins insert themselves, transmitting their deformity to the rest of the system through contact. Prions are agents of disorganization in a morphological continuum maintained by contiguity.

Physiological modification of the condition of protein regimes is brought about by electromagnetic forces. A palisade in the liquid crystal of a cell membrane takes up a spatial orientation governed by electro-magnetic forces that surround and penetrate the cell. As the proteins assume a direction under the influence of these forces they determine and stabilize their neighbors, rather like blades of grass in a meadow that capture a gust of wind and bend to its thrust. The direction in which the molecules point has been referred to as their "director" or "attractor." If the direction shifts as the wind changes, so the proteins bend docilely toward the new pole of attraction and alter their orientation, each one lending support to the next one. When a change of phase occurs that solidifies the crystal, the meadow becomes still and the blades of grass remain prisoners of the last gust. This solidarity of millions of ordered molecules makes up a "coherent regime."

Electric or magnetic forces coherently excite the developing organism and form a "field" that both orients its molecular cohorts and is expressed by them. The hydra is an aquatic animal consisting of little more than a crown-shaped mouth at one extremity and a foot at the other. The mouth end is electrically positive and the foot end is negative. By passing weak currents through the budding creature, its polarity can be inverted and the mouth switches to the other end. Eggs, too, have an electrical polarity. Small oscillating electric currents make their first appearance in the fertilized egg. They flow through the embryo, discharging here and there, provoking cells to jump and scurry and produce folds and convulsions that lead to new patterns and yet more flows and contortions. The dynamic totality of these vector forces is referred to as a "morphogenetic field," and it is through this field that little by little the form of the organism is constructed. Liquid crystals contain and express this field. In the nuclei of cells caught up in this minute living magma, the genes switch themselves on and off to provide material to the arcane furnaces.

It is not the genes that elicit nascent form, but the nascent form that selects the genes and recruits them for its program.

Crystals in general—and liquid crystals in particular—refract incident light and, under the right conditions, generate a range of brilliant colors. As the orientation of the molecules changes synchronously in the flexible strata, the reflected lights change in intensity and color. The English-Chinese researcher Mae Wan-Ho, along with her fellow microscopist M. Lawrence, had the skillfulness to discover a new imaging technique based on the refraction of liquid crystals. Watching on a screen, she exclaimed jubilantly: "We can see all the colors of the rainbow in a living, crawling, first instar *Drosophila* larva.... Life is all the colors of the rainbow in a worm!"

The larva, only one millimeter long, comes into focus on a color television monitor "as though straight out of a dream." As it crawls along it weaves its head from side to side, its jaw muscles flashing in blue and orange stripes against a magenta background. The bands of segmented muscles switch from brilliant turquoise to a bright vermilion, tracing luminous wavelets along its waving body. As the muscles contract, the walls of the body vary in color from magenta to purple, with iridescent hues of green, orange and yellow.

The dance of lights that accompanies and marks the life and development of the fly larva is orchestrated by electric currents that orient the liquid crystals, which take in the currents and re-express them. Exposure of the fly embryo to weak magnetic fields sets off aberrations in the larva's segmentation. Magnetic fields are known to produce spiral waves and turbulence in the electric fields of liquid crystals. Changes in the arrangement of molecules are transmitted at a distance thanks to the solidarity typical of coherent regimes.

In imposing cohesion, the coherent state nevertheless allows a certain local freedom. Mae Wan-Ho compares it to "a symphony orchestra or a grand ballet, or better, yet, a jazz band where every individual is doing his or her own thing but yet is in tune or in step with the whole." This gives us another metaphor for the relationship between the parts and the whole.

Referring to the three states of matter, we took as our example water, which can be ice, liquid or vapor. Ice can be glassy or crystalline, as in snowflakes, but water in the liquid state seems to be devoid of structure, of any memory, a symbol of purity and innocence.

St. Francis wrote: "Laudato si', mi Signore, per sor'aqua la quale è molto utile, et humile et preziosa e casta." ("Be praised, my Lord, for our sister water, who is very useful, and humble, and precious, and chaste.")

From the physicist's perspective H_2O, the water molecule, is rather like a scribing compass, with an oxygen atom at the vertex and a hydrogen atom at the ends of each leg. This makes the molecule an electric dipole; as such, it takes on a direction in an external electric field. The water of the physicist is not as gentle as the water of St. Francis; it has a predisposition to order and obedience because its molecules are ever on the alert, like tiny antennae. When these molecules exchange messages, they are able to bring their oscillations into phase and amplify them. They thereby enter into a new fundamental state, a coherent regime. The temporal rhythm of the oscillations is transformed into a spatial framework. (G. Preparata)

Compared to those of liquid crystals, coherent regimes in water are much less extensive—of the order of 0.1 micron, or one-tenth the size of a bacterial cell. These "micro-cells" of ordered water swim about in wild water, niches of order in an ocean of chaos. They are models or precursors of life's order. In one sense coherent water has a memory, which is its very order, its field. (E. Del Giudice)

Now that we know there are conditions for producing order that have an autonomous existence and a memory of their own, even in the humblest and chastest of liquids, let us leave the magic of water and turn back to biology's morphogenetic fields.

The chief interest in biology's morphogenetic field lies in the fact that it maintains and transmits itself autonomously apart from the genetic endowment archived in DNA. During development the two systems influence each other, the field awakening the genes and the genes supplying the field with the material it needs. The big news in late twentieth-century biology was the recognition of a second hereditary torrent rushing alongside the stately flow of the DNA channel. It will be no easy task to build up topographic maps of developing fields because of their three-dimensional nature, their turbulence and their continual transformations. But it may be those very maps that contain the secret that distinguishes between the fly and the whale, between the mussel and the horse.

I can well imagine someone asking me why, having taken up macroscopic hypotheses and metaphorical morphic fields, which defy the logic of cause and effect, and even the winging arrow of time, I have in these last two chapters fallen back on a molecular interpretation, if only to attribute a newfound dignity to proteins. I have asked myself the same question. To be sure, there was no premeditation in the move. I simply want to comment that the bond that allows proteins to tell each other about their state and to transmit it through heredity is not of a mechanical or causal nature. It is a solidarity of orientation, a "morphic resonance" (Rupert Sheldrake), whereby every element in a coherent system is subjected to a force and itself expresses that force. Natural systems have an inherent collective memory. This imparts coherence to molecules, crystals, cells, termite colonies, dragonfly swarms, flights of birds, and the myths of mankind. As Sheldrake puts it: "Things are as they are because they were as they were."

CHAPTER XV

THE LEAF INSECT BEFORE THE LEAF

Many animal species fade so completely into the background where they dwell that it is difficult to see them, and we notice them from time to time only when they move. On snow-covered hills, white partridges, lemmings and polar foxes have plumage or snow-white coats that defy the observer's powers of detection. But when the snows melt with the arrival of spring, the white camouflage disappears and the animals put on the dun-colored coat of the warm season and of the land itself. In the desert, animals mostly have the color of the sand or rocks. The eagle owls, isabel goatsuckers and lark, among birds; the gazelles, gerbils and mice among mammals; and the horned viper, rattlesnakes and many lizards among reptiles—all take the Franciscan habit and assume the hues of humility. When the rarest of rainfalls brings the beauty of green to the desert, caterpillars and green insects suddenly appear on the tufts of grass to mimic the unexpected chlorophyll.

Leaves, fresh or dry, are a favorite place for mimetic armies of living things. Some of the tiny inhabitants of foliage rely on a green camouflage, but the cleverest among them mimic even the ribs and outer shape of the leaves. They become leaves among leaves. To Goethe's "Alles ist Blatt," they respond: "We're leaves, too." With its wings folded the *Kal-*

lima butterfly resembles a perfect leaf surface, but when it spreads its wings in flight it reveals on its upper side the garish colors and the joyful eyes of Madame Butterfly herself. Perfect mimicry is seen in *Phyllium* butterflies, certain long-antennaed locusts, and a tropical mantis from America. The mantis has lamellar expansions on its feet and body, giving it the appearance of bundles of leaves. Locusts, in order to lend verisimilitude to their make-up, take on erosion-like patches that make their bodies look as though they were being attacked by mold.

Usually leaf imitators copy dry leaves or broken pieces of them, and they hide themselves in the sober carpet of fall, in which all leaves resemble one another in their return to the earth. There is a small South American fish that looks like a leaf and lets itself float on the surface of the water, just like the dead leaves that the river carries down to its estuary.

Mimicry is among the arguments most eagerly adopted by neo-Darwinists, because it supposedly demonstrates animal artfulness and provides support for utility as opposed to beauty. Mimetic animals renounce themselves almost as a condition of being there at all. They represent the opposite of Portmann's self-presentation. Mimetics cancel themselves out. It is understandable that this serves them some purpose in ensuring their survival, but an animal also needs to be recognized, to attract or scare off, to express itself. As far as I am aware, no one has demonstrated that mimetics are mutants of their clown brethren; if anything, they offer the most obvious evidence of modified appearance without a change in DNA—without a change in identity. Animals that take on a white cover when the snows fall and turn brown when the snow thaws in spring tell us that their camouflage is the child of the cold and not of changes in their genes. The Himalayan rabbit has a thick white coat. If a letter of the alphabet is shaved out of the coat and the rabbit is transferred to a warm environment, the coat grows again at this point but with the letter filled with dark fur. The two *Vanessa* moths that August Weismann asked in his will to be included in the last photograph of himself are genetically the same, but one has reddish wings and the other yellow wings—the difference being ascribable to a direct effect of the seasons.

Many animals that resort to mimetic coloring—the chameleon provides the prime example—do so by changing their basic coloring, sometimes after several days, at other times all of a sudden. The eared chameleon of western Africa changes from dark green to light green, while Fischer's chameleon alters its brown hue from dark to light. Cephalopods—the cuttlefish, inkfish, and octopus—change color suddenly as they shift from one background to another. If a spider crab alights on a yellow *Ranunculus*, it turns bright yellow itself; if it is placed on a white violet, its color fades.

None of these color changes have anything to do with genetic variations. They are physiological changes that animals in chiaroscuro effect by acting on surface cells that contain a granular black pigment, melanin. When animals disperse the pigment they turn dark, and when they centralize it they take on a paler hue. Other animals have colored pigments.

For all the artfulness we can ascribe to mimetic animals, their equivocal, chameleon-like cleverness has very little connection with genetic variation. The much-discussed experiments on peppered moths only suggest short-term genetic adaptation, and it is so very sad that they have become the paradigm of a phenomenon that is, if anything, the most glaring example of how animals can change color and appearance without undertaking the long, precarious, lethal genetic journey.

Interpreting cases of camouflage that involve the copying of elaborate forms is more intriguing. Some animals can flatten themselves against their background, others roll up into balls, others stretch out; still others astound us by the precise forms that they are able to copy, stably and in finest detail.

A strange order of insects, close to the coleopterans, has been given the name of phasmids ("phantoms"), indicating their ability to remain invisible to neighbors by imitating the forms and colors of the trees on which they alight. Stick insects and leaf insects belong to this order. These astute animals are taken as examples of mimetic adaptation, but they proved embarrassing when paleontologists started following their fossil traces and found them where they were not supposed to be. How these ghost insects—these incredible mimickers of leaves and sticks—ever came into the world remains a mystery, an unsolved scientific detective story. The reigning utilitarian interpretation would have it that these insects, before they were like leaf surfaces or dry sticks, got mixed up with leaves and twigs and, through mutation after mutation, came to resemble their background, until they arrived at the point of becoming the perfect models we now see (to the point that we don't see them!) on plants. Unfortunately for them and for the theory, these artful imitators derive no benefit from their mimetic capacity and are prey for their enemies, which have no difficulty in detecting them. But the most unforeseeable surprises have come from paleontology. The oldest phasmid fossils (they go back in Baltic amber to the Tertiary—i.e., about 50 million years ago) look identical to present-day species, showing that no gradations have occurred. It is thought that those phasmids originated from Chresmodids of the Upper Jurassic in Germany, fossils of which are encountered in deposits dating back some 150 million years. But the oldest fossils of stick or leaf insects (protophasmids) go back to even remoter periods, in the Permian (250 million years ago, in the Paleozoic). One might argue that these insects completed the process of imitating leaves at an extremely gradual rate beginning at a still earlier time. Yet things do not work out this way. Plants with flowers and leaves (phanerogams and *Latifoliae*) appeared no earlier than the Cretaceous—in other words about 100 million years ago, long after the first protophasmids. This chronological anomaly places the imitators earlier in time than the objects of the imitation, leaving entomologists and paleontologists disconcerted. In his *Great Book of Nature* (Hoepli, 1954), Fritz Kahn asserts that there must have been a mistake somewhere. How could the leaf insect be older than the *Latifoliae* on which it has modeled itself? I once saw a picture of a leaf insect in a reputable journal, in which the commentator attributed to the artful imitator an ability to "prophesy" something that originated millions of years later!

Leave out mimicry. We have to recognize that phasmids have a tendency to laminate themselves like leaves or elongate themselves like sticks, and they had this tendency prior to and independent of the plants. Who knows? Perhaps phasmids mistake the plants' leaves and sticks and for themselves?

The entomologists I have consulted prefer to gloss over the phasmids. The Chresmodids (150 million years ago) have been disqualified as their dubious forebears, and the protophasmids have been excluded from their phylogeny altogether. The exclusion does not change the general picture, however. There were leaf insects and stick insects—"phantasm" insects—over 100 million years before there were leaves or sticks. "It is as though," wrote R. Lewontin, "railroad lines and stations, and their roofs and signal boxes, were discovered a thousand years before the train was invented."

Lewontin uses this metaphor to comment on another, even more incredible discovery, which gives the lie to the *idée reçue* that every organ of every organism represents the evolutionary "solution" to a survival problem. He refers to an article by C. Labandeira and J. Sepkoski appearing in *Science* July 18, 1993. Their research dealt with the buccal (mouth) apparatuses of insects. In their tininess these usually escape observation; but if they were the size of, say, a dog's muzzle, we would see horrendous mouths of monsters or machines—even though their food was offered to them by the grace of flowers. The insect's mouth is fourfold and is always composed of the same parts: upper lip, upper and lower jaws, and lower lip—among which is sometimes a prominent pharynx. In masticating insects—locusts, beetles and roaches—the upper jaw and the lower lip are robust pincers consisting of various elements—hinge, stipes, outer lobe, inner lobe, palpiger, and long, articulated palps. In flies, what predominates is the lower lip, which is equipped with substantial labella that are able to suck up any liquid, and which is sometimes found elongated and hardened for penetration. In the lambent bee mouth, designed to lick up nectar, the lower jaw and lip lobes are elongated and promi-nent, forming a spatula. In butterflies, the lobes are greatly elongated and unite to form a rolled-up proboscis. In some rhynchota, the lower jaws are transformed into four hairy styli and are located on the lower lip, which has become an elongated beak providing a sheath for the four dagger-like processes.

In order to reconstruct the history of these buccal apparatuses, La-bandeira and Sepkoski examined the fossils of over 1,200 families of insects. The expectation was that these would have begun to diversify following the expansion of flowering plants, which supposedly gave rise to the "problems" to which the insects were expected to find their "solu-tions." Yet the expansion of flowering plants took place only in the last 100 million years, whereas insects began their increased diversification 320 million years ago, and the number of insect families increased ex-ponentially during the first 150 million years of their existence. Once flowering plants began to make their appearance, the increase in number of insect varieties slowed almost to a standstill. Thus some 85 percent of insect families were there millions of years before the first flowers burst out of their buds.

The insect establishes a predominantly buccal relationship with its flower. When Labandeira and Sepkoski studied the structural parts of the buccal apparatuses of fossil and living insects in more depth, they concluded that practically all (more than 95%) of the conformations of those parts were present before the appearance of flowering plants.

How did it happen, then, that these complex and delicate appara-tuses existed millions and millions of years before they had a job to do? They did not originate as a solution to one problem or a thousand prob-lems, nor were they the result of natural selection. What other picture can one propose to make sense of this reversal of history, this turning backwards of time?

It seems to me that all this is a point in favor of the idea that liv-ing reality is an exercise in the possible—an exercise that, as millions of years succeeded one another in this vague and pointless pastime called reality, pulled out all the stops, ready to make itself useful or leave itself unutilized, making a display of the marvelous adaptability of living things. But those insects still had to feed themselves, for hundreds of millions of years, among mosses, ferns, cicadas and conifers, in the empire of great reptiles. And eat they did: They had knives and forks and spoons, richer and more sumptuous than their lean and wooden fare required. And when succulently smelling food and gaily colored table linen arrived on the scene—the great feast—practically the entire buccal cutlery was ready for use, and the vanity of the insects was at long last satisfied. Even today, I would be prepared to wager, many of the parts of the buccal apparatus could change, without the slightest detriment to the owner; though, with so many things to take up their time, their imagination has recently been reduced. Flowering plants have invaded the entire morphological space, everywhere they have found mouths to feed, as though the plants were restaurant keepers preparing the table with viands suited to the habits and preferences of their customers.

CHAPTER XVI

THE HIDDEN ROOTS OF LIVING FORMS

"Today, nearly half a century after the publication of the Encyclical *Humani Generis* (1950), new knowledge leads to the recognition of the theory of evolution as more than a hypothesis." Thus said Pope John Paul II in his address to the Plenary Assembly of the Pontifical Academy of Sciences (1996).

I think that any phenomenon in nature becomes a matter for science only when it is underpinned by a solid theory. The eclipse of the Moon is a matter for myth as long as it represents a dragon devouring the queen of the night, or Zeus devouring Themis. It becomes a matter for science when we regard it astronomically, as the passage of the moon through the earth's shadow. The astronomical theory makes possible countless verifiable predictions and computations. A theory, as the Pope pointed out, is constantly tested by comparing it with facts. "Where it can no longer explain [the facts], it shows its limitations and unsuitability. It must then be rethought."

Darwinian theory is based on the notion that living forms gradually turn into others, or gradually emerge in divergent forms, when they become separated into different environments. No one has ever witnessed such phenomena, but evolutionary theory predicts certain outcomes of these processes. For example, the theory postulates lots of "intermediate forms" in the fossil record, forms that are assumed to have been branching outward for millions of years. In the case of molecules, the theory predicts a series of adaptive changes (mutations) that supposedly account for morphological differences. The latter probably constitute the "new knowledge" to which the Pope alluded.

I would like to inform the Holy Father, however, that he has been misinformed. Molecular biology has indeed certified the unity of living things, but apart from this it has disappointed the central expectation of the theory: the molecular explanation of their diversity. Organisms differ widely morphologically, yet they are cytologically similar and biochemically almost alike. True, the more distantly related two living forms are, the greater their molecular differences—but these differences have little to do with their forms. No molecule in the clam is the least bit more marine or more mollusk-like than its equivalent in the horse.

No bat molecule is more winged or more clinging than a whale molecule. The changes we are talking about are the outcome of a neutral history in which messages have been worn down without having their meanings altered. Morphological differences repose on another divan altogether.

For billions of years life remained microscopic and single-celled. In a geologically short period of time, as if by a spell, the major forms of animal life that subsequently populated the earth appeared side by side, in all their morphological "types," or phyla. It is calculated that this happened half a billion years ago, at the beginning of the Cambrian. Nematodes (roundworms), annelids (earthworms), arthropods (insects and crustaceans), echinoderms (sea urchins and starfish), chordates (our phylum) and many others emerged at the beginning of the Cambrian. Below that there is no fossil whatever that could have generated them. This "explosion of life" was demonstrated by Roderick Murchison (1830-40) some twenty or thirty years before *The Origin of Species* (1859), and Darwin, in all honesty, had no explanation to offer. The latter acknowledged that "several of the main divisions of the animal kingdom suddenly appear" in the Cambrian, and that "to the question why we do not find rich fossiliferous deposits" containing their presumed pre-Cambrian ancestors "I can give no satisfactory answer." Nor, a century and a half later, has an answer been forthcoming.

This shows that animal life did not emerge at the mouth of some delta, as the result of a gradual transformation and diversification of forms—the Darwinian mechanism—but rather as an immense improvisation, in a scenario more reminiscent of Cuvier's catastrophism.

At the time when multicellular life originated, the Earth was populated by as many species as were there subsequently, and since then it has witnessed in amazement occasional swarms of substitutions. The living world formed from "concerted explosions" over diffuse areas through the "revelation" of unexpressed forms, not because of local inventions and their geographical dissemination.

The explosion of types is not simply an insignificant, unresolved charade in the Sphinx's book of riddles. It is exactly the opposite of what Darwin's gradualist mechanism predicted regarding the origin of animal forms. Yet the evolutionists do not seem to have been unduly troubled by this. Their theory of adaptation will adapt to anything.

The neo-Darwinists or "selectionists" who put forward the theory were geneticists and biochemists, people largely innocent of—and scarcely interested in—the history of the Earth. Their world was a laboratory world, a test-tube world, and they were more interested in manipulating it than in understanding it. As often happens in science, they thought their little laboratory tricks would turn out to be the means employed by nature's intelligence in constructing the biosphere—a biosphere that they judged to be redundant. It was as though they were boasting that they made no distinction between a fir tree and a larch, or between a roebuck and an ibex. Because chance plays a decisive role in their theory, it was of little consequence how things actually happened—like numbers in a lottery. An unkind joke has it that a scientist once made a bet with a shepherd on the number of sheep in the flock, with the winner to get a sheep. The scientist guessed correctly, and with great satisfaction took up an animal, only to hear the shepherd say, "So you're a molecular biologist, then?" "But how did you guess?" replied the scientist. "Never mind," said the shepherd, "just put my dog down."

The vertebrate type (actually, subphylum) is divided into classes. Among these are the jawless fish (lampreys), cartilaginous fish (sharks), bony fish, amphibians, reptiles, birds and mammals. Every attempt to demonstrate the derivation of recent forms from more ancient ones, and to find intermediates, has met with extremely poor results. The much sought-after "missing links" are still missing.

When the biochemists embraced evolution in the 1960s, they forthwith devoted themselves to that irresistible passion of the evolutionists, the attempt to construct a grand genealogical Tree of Life, this time on a molecular basis. In 1969, Margaret O. Dayhoff affirmed that "a tree constructed in this way has the same topology as that derived from conventional morphological considerations." Yet the tree was full of contradictions. I myself, together with my wife Isabella, worked on the same material as that on which Dayhoff based her statement—i.e., variations in the 100 amino acid cytochrome c molecule. We examined published protein sequences in the lamprey, shark, frog, tortoise, emu, horse, macaque monkey, and rattlesnake. We concluded that the data were compatible with the hypothesis (we called it the null hypothesis) that the cytochromes of all classes derived synchronously from one and the same molecular progenitor, and that there was no order of precedence among them. (*Rivista di Biologia*, 1987) Mammals, tortoises, amphibians, sharks and lampreys turned out to be separated brethren, none arriving on the scene before, or ancestral to, any other. One or two perplexities remained. The primates (our order) had covered a long stretch of the journey in common with the snakes; the birds were related in some way with the tortoises, and the shark with the lamprey. Other authors had noted the same anomalies. The curious kinship of Man with the Serpent has a Biblical ring about it. The two appear suddenly as the offspring of an angel's egg. The Serpent, deprived of limbs, is condemned to creep in the same episode in which Man, by reason of his hot blood, is condemned to sweat and Woman is condemned to suffer in childbirth. And all three lose their wings.

The most embarrassing thing to emerge from these molecular studies was that all the differences observable among homologous proteins in the various classes were "neutral"—they had no influence on function or form. Chicken cytochrome c functioned perfectly well in spinach. No one had encountered mutations to which one could assign the differences between the human and the monkey, or between the chicken and spinach.

If we return to the documentary evidence offered by paleontology—the fossils—we note that modern mammals (*Eutheria*) provide the most impressive example of concerted explosion. More than twenty orders of mammals emerged almost at the same time, at the beginning of the Cenozoic—i.e., some 65 million years ago. The first to emerge among the fossil remains were the primates and the insectivores. These were followed immediately by the rodents, carnivores and ungulates, with cetaceans bringing up the rear along with the sirens, proboscidates and herbivores. Practically all of these made their appearance within the dozen or so million years of the Paleocene—i.e., more than 470 million years after animal forms first originated. Paleontologists have considerable difficulty unearthing intermediate forms between the reptiles and the mammals, and they cannot even imagine gradations among some of the orders of mammals, which embrace living beings so different as the bat and the whale, the mouse and the elephant.

No semi-bat has ever been found, and no one has ever thought up a good common ancestor for the fantastic bestiary of the mammalian orders. The genealogical tree that shows forms gradually varying and diverging—the "fact" for which Darwinism proposed its revolutionary explanation—is nowhere to be found. What we find instead are bushes with no roots, popping up from nowhere, prodigal with different sorts of flowers that resemble one another but do not know each other. If we must find a multiparous mother for them she might be pictured as a tiny, elusive something that harbors everything and expresses almost nothing, a shy larva that reproduces its pallid next-to-nothingness in expectation of one day fluttering butterfly-like in multitudinous multicolored forms.

Pierre-Paul Grassé pictured such evanescent Great Mothers as archaic forms that were generalized and unadapted, almost as fecund embryos. Even if we were to find them, they would tell us nothing about their destiny, a destiny that might reveal itself only after prodigious deliveries, evoked by some unidentified attractors. Grassé compares them to an underground rhizome, like strawberry stolons from which rosettes of little leaves spring forth from time to time. The submerged stolon propagates uninjured, because it never expresses itself; it is preserved because it hides; and when it reaches the nodes from which forms irradiate, it finds in these its moment of glory and its death.

Grassé likens the explosion of related forms to a whorl of twigs expanding from the node of a stele (the archaic mother). The convergence of the offshoots does not mean that all types are the offspring of a single excruciating labor. Archaic sister larvae could be dispersed over a vast area and flutter here and there in imago form until Time one day summons them. Think of the bamboo, and how after many green years these plants all flower together, at all latitudes and altitudes in the world, then die.

At the time Grassé was formulating his model, Léon Croizat—roaming the continents and searching for plants and animals, living and fossilized—was developing his own theory of "panbiogeography." (See Chapter X above.) Croizat envisioned a phase in which an archaic form scurries about in the vastness of its territory ("mobilism"). Subsequently it breaks up and settles in separate areas ("immobilism"), which may become further isolated from each other by the emergence of natural barriers. The subgroups thus isolated—geographically or ecologically—begin developing autonomously and give rise to distinct forms which are similar to one another ("vicariant") but unknown to each other, having become separated at an undifferentiated, archaic time when they could not yet be distinguished. Thus, in the savannahs of the southern continents, the ostrich, emu, nandù, cassowary and kiwi appeared synchronously and in scattered fashion. Something similar happened with the orders of mammals: carnivores, chiropters, cetaceans, primates, etc. Shared internal forces and general rules of development made them all brethren, while varieties of habitats made them brethren differing in form.

With his patient explorations and observations Croizat traced lines that linked the various expressions of one and the same order. The resulting tracks, which crossed continents and impassable oceans, were the trails traveled by archaic mothers before the continents began drifting, over a land that is now lost, in another history and another geography.

Explorers of the forms of human myths, Giorgio de Santillana and Hertha von Dechend (co-authors of *Hamlet's Mill*), followed up clues in the Hamlet legend. They did so in a whirlwind tour of Shakespeare and *Saxo Grammaticus*, of the *Edda* and the *Kalevala*, of the *Odyssey* and the *Epic of Gilgamesh*, journeying from Mesopotamia to Iceland and from Polynesia to pre-Columbian Mexico. The same myth, with the same elements, has been retold in different tongues by people who never knew each other. The myth was not born in a single region from which it spread while varying gradually to different areas. All versions are in their own way original. These writers concluded that "the original life of the mind opened up a way for itself in the dark, spreading its roots and tendrils in the depths until such time as the living plant emerged into the light under different skies. Half a world away it is possible to discover the same journey of the mind."

All these representations falsify the idea that adult forms change into derived forms gradually as they migrate elsewhere, accumulating copying errors on the way. The tiny rodent gradually becoming a hippo and then a whale, as the Darwinian vulgate would have it, follows a superficial trajectory connecting adult forms via undiscoverable intermediaries. Ernst Haeckel lined up a series of imaginary animal forms in which each was the fetus of the form that came after, and he suggested that this trajectory was the path followed by evolution. He called this his Biogenetic Law: "Ontogeny recapitulates phylogeny."

Grassé's rhizomes, like Croizat's tracks, hark back to Weismann's germinal lines. Weismann maintained that the chick is not the offspring of the brood but that chick and brood are both lateral branches of the same germinal line. Grassé denies any branching between classes of the same phylum or between orders of the same class; instead, he sees them as siblings radiating out from the same stolon. The Turin evolutionists, of the school of Rosa and Croizat, postulate parallel evolutions and rely on a similar representation.

Between related forms there is no other connection beyond the provisions for the journey that they possess in common, and that have been transmitted to them by the same vagrant mother. Each one makes use of the provisions as it pleases, extracting from this maternal totipotency its own special design.

The condition here described can be looked at in another way. We can imagine that between groups separated by insurmountable barriers there nevertheless remains some form of communication at a distance, a subtle telepathy, that somehow brings into their morphological development what Rupert Sheldrake has called "morphic resonance." In the heavens, two galaxies 200,000 light years distant apparently exchange with each other some influence remembered from a near collision 200 million years ago.

The line linking the brood and the chick is something outside the timeline of life: Brood and chick are contemporaries and siblings. But the same is true of the lines linking the amphibian and the reptile, the whale and the bat. These are outside history and run their course in a time that has nothing to do with our growing old.

Siblings that originate as a result of radiation are contemporaneous and coeval. They enter into time when their development in the world begins. The laws they express are outside life; they prescribe forms for history, and history selects these from the vast but finite inventory of the possible.

The Darwinian model is a mundane history where everything changes but nothing ever happens. It is all passion for change, for becoming, for growth—in other words, for that voyage of entropy that leads eventually to death. Just as the larva metamorphoses into the insect, or the embryo into the adult, so living forms are supposed to transform one into the other, following the branches of the great tree of life.

Actually, these metamorphoses of living things are only appearances. When the caterpillar leaves the stage to the butterfly (the imago), it does not transform its sluggish feet into delicate flying pages. Enclosed in the pupa, the caterpillar simply decomposes, and only certain groups of embryonic cells remain alive—the imaginal disks, which form the butterfly ex novo, like a sibling taking the place of the sacrificial victim.

In ordinary conversation, blood is a metaphor for the continuation of the race or lineage. According to the most recent embryological thinking, blood is a tissue that does not continue. Instead, its cells are produced by extra-embryonic "blood islands," which put into circulation special cells that establish bridgeheads in the blood-forming (hematopoietic) tissues—liver, marrow, spleen, and thymus. These stem cells, as they are called, are pluripotential, and they persist throughout adulthood, especially in the bone marrow. These later diverge into three different derived stem lines, which produce red blood corpuscles and two different kinds of white corpuscles. Specific stem cell lines are present in other tissues as well, such as the liver and the brain, where they are ready to mature into adult cells. In an individual organism the stem lines perpetuate the conditions in which classes or orders make their appearance from Grassé's rhizomes. The question about what persuades a pluripotential stem line to turn into red or white blood cells is the same as the question about how a pluripotential "rhizome" became a mouse, or a whale, or a bat.

Stem cell lines contain not only the instructions for the materials, but also the architectonic maps to assemble the forms. Beside the universal DNA, there is a complex of regimes of physical coherence, "something rich and strange," which produces diversity.

The mystery of evolution is rather like a humble procession of pilgrims who decide to found various cities here and there along the way. Each city develops its own forms and habits, like the orders of mammals. Though united in their founding, these may have paws for trotting in the plains, wings for fluttering about in the night air, fins for diving in the seas, or arms for climbing trees. Or minds for philosophizing.

RISING FROM THE FLATLAND

Why, after two centuries of evolution, are we still engaged in discussing the meaning of the term itself? If Evolution is a field of knowledge like Geography, History or Geology, we should not have to ask about its occurrence. Do we ever ask if Geography exists?

All our quarrels are about hidden agendas. Speaking of evolution, one irresistibly thinks of Progress, from simple to complex, from primordial to refined. "It's not a matter of romantic progress," object the experts, "evolution is an issue of adaptation." "Adaptation? What do you mean—that old story that the fittest, defined as those which survive better, survive better?" (Waddington) Evolution, we are eventually told, is only Change—change in gene frequency, and that's all.

So we have a theory of progress and/or adaptation from which progress and adaptation have been removed. There is still, however, room for disagreement: Are the changes gradual or abrupt? Are they transformations or replacements? "Transformations" is the reply of the geneticist, "we read them in molecules." "Replacements," some geologists object, "the links are missing, and your DNA changes are merely neutral. They don't produce evolution."

One last detail. Abrupt changes cannot occur without models, designs, concealed potentialities, laws of growth—as in ontogenetic development. These ideas hold perhaps the key to evolution, the last hope. "But they do not belong to the accepted paradigm. They are not scientific," object the reductionists. So we face an alternative. Either we anchor our ship to a desperate Science paradigm that offers no possible exit; or we propose a change in paradigm, the emergence of a new space dimension in Science, from which we can reassess the problem of evolution.

It is time to rise from the Neo-Darwinian flatland to the realm of forms.

BIBLIOGRAPHY

Augros, R. M., and G. N. Stanciu. 1987. *The New Biology: Discovering the Wisdom in Nature*. Boston: New Science Library.

Balon, E. K. 1986. "Saltatory Ontogeny and Evolution," *Rivista di Biologia/Biology Forum* 79: 151–190.

Barbieri, M. 1985. *La Teoria Semantica dell'Evoluzione*. Torino: Boringhieri.

Baskin, J. 1988. "Leon Croizat: Out of nothing nothing comes. Interview with Leon Croizat (1974)," *Rivista di Biologia/Biology Forum* 81: 589–611.

Beloussov, L. V. 1998. *The Dynamic Architecture of a Developing Organism*. Dordrecht: Kluwer Academic Publishers.

Bird, W. R. 1987. *The Origin of Species Revisited: The Theories of Evolution and of Abrupt Appearance*. 2 volumes. New York: Philosophical Library.

Bowden, M. 1988. *Ape-Men—Fact or Fallacy?* 2nd edition. London: Sovereign Publications.

Buffon, G. 1797–1807. *Barr's Buffon: Buffon's Natural History; containing a theory of the Earth*. 10 volumes. London: J. S. Barr.

Chagas, C. (editor). 1983. *Recent Advances in the Evolution of Primates*. Vatican City: Pontificia Academia Scientiarum.

Craw, R., and G. Sermonti (editors). 1988. Special issue on "Pan-biogeography, Space—Time—Form," *Rivista di Biologia/Biology Forum*, 81(4).

Croizat, L. 1964. *Space, Time, Form: The Biological Synthesis*. Caracas: Published by the author.

Curtis, H. 1980. *Biologia*. 2a edizione. Bologna: Zanichelli.

Dawkins, R. 1986. *The Blind Watchmaker: Why the evidence of evolution reveals a world without design*. New York: W. W. Norton & Company.

Dayhoff, M. O. 1969. "Computer Analysis of Protein Evolution," *Scientific American* 221: 86–95.

Dayhoff, M. O. 1978. *Atlas of Protein Sequence and Structure*. Silver Spring, MD: National Biomedical Research Foundation.

Denton, M. 1985. *Evolution: A Theory in Crisis*. Bethesda, MD: Adler & Adler.

Dewar, D. 1957. *The Transformist Illusion*. Nashville, TN: Dehoff Publications.

Dickerson, R. E. 1972. "The Structure and History of an Ancient Protein," *Scientific American* 226: 58–72.

Dike, C. 1988. *The Evolutionary Dynamics of Complex Systems: A Study of Biosocial Complexity*. New York: Oxford University Press.

Driesch, H. 1892. "The Potency of the First Two Cleavage Cells in the Development of Echinoderms," reprinted in B. H. Willier and J. M. Oppenheimer (editors), *Foundations of Experimental Embryology*, 1964. Englewood Cliffs, NJ: Prentice Hall.

Fantappié, L. 1941. *Principi di una teoria unitaria del mondo fisico e biologico*. Rome: Di Renzo.

Florenskij, P. 1914. *Le Porte Regali: Saggio sull'Icona (a cura di E.Zolla)*. Reprinted 1977. Milan: Adelphi.

Goodwin, B. 1994. *How the Leopard Changed its Spots: The Evolution of Complexity*. New York: Charles Scribner's Sons.

Goodwin, B., A. Sibatani, and G. Webster (editors). 1989. *Dynamic Structures in Biology*. Edinburgh: Edinburgh University Press.

Gottlieb, G. 1992. *Individual Development & Evolution: The Genesis of Novel Behavior*. New York: Oxford University Press.

Gould, S. J. 1977. *Ontogeny and Phylogeny*. Cambridge, MA: Harvard University Press.

Grassé, P.-P. 1977. *Evolution of Living Organisms: Evidence for a New Theory of Transformation*. New York: Academic Press.

Greene, J. C. 1959. *The Death of Adam: Evolution and its Impact on Western Thought*. Ames, IA: University of Iowa Press.

Hammen, L. van der. 1988. *Unfoldment and Manifestation. Seven Essays on Evolution and Classification*. The Hague: SPB Academic Publishing.

Imanishi, K. 1984. "Conclusion to my Study on Evolutionary Theory," *Journal of Social and Biological Structures* 7: 357–398.

Ho, M.-W. 1993. *The Rainbow and the Worm: The Physics of Organisms*. Singapore: World Scientific.

Ho, M.-W., and S. Fox (editors). 1988. *Evolutionary Processes and Metaphors*. Chichester: John Wiley and Sons.

Hooker, J. D. 1853. *Flora Novae Zelandae*. London: Lovell Reeve.

Huxley, J. 1942. *Evolution: The Modern Synthesis*. London: Allen & Unwin.

Ikeda, K. 1988. *What is Structuralist Biology?* In Japanese. Kamey Sya.

Jacob, F. 1974. *The Logic of Life: A History of Heredity*. Translated by B. E. Spillmann. New York: Pantheon Books.

Jacob, F. 1977. "Evolution and Tinkering," *Science* 196: 1161–1166.

Kaufmann, S. A. 1993. *The Origins of Order: Self-Organization and Selection in Evolution*. Oxford: Oxford University Press.

Kettlewell, H. B. D. 1973. *The Evolution of Melanism*. Oxford: Clarendon Press.

Kimura, M., and T. Ohta. 1971. *Theoretical Aspects of Population Genetics*. Princeton: Princeton University Press.

Kimura, M. 1979. "The Neutral Theory of Molecular Evolution," *Scientific American* 241: 98–126.

Labandeira, C. C., and J. J. Sepkoski Jr. 1993. "Insect Diversity in the Fossil Record," *Science* 261: 310–315.

Lambert, D. M., and H. G. Spencer (editors). 1995. *Speciation and the Recognition Concept: Theory and Application.* Baltimore: John Hopkins University Press.

Lima-de-Faria, A. 1988. *Evoluzione senza Selezione: Forma e Funzione per Autoevoluzione.* Genoa: Nova Scripta.

Lima-de-Faria, A. 1995. *Biological Periodicity: Its Molecular Mechanism and Evolutionary Implications.* Greenwich, CT: JAI Press.

Løvtrup, S. 1987. *Darwinism: The Refutation of a Myth.* London: Croom Helm.

Løvtrup, S. 1982. "Le quattro teorie dell'Evoluzione," *Rivista di Biologia/Biology Forum* 75: (1, 2, 3, 4).

Lovenstam, H. A., and S. Weiner. 1989. *On Biomineralization.* New York: Oxford University Press.

Macbeth, N. 1971. *Darwin Retried: An Appeal to Reason.* New York: Dell Publishing.

Mandelbrot, B. B. 1983. *The Fractal Geometry of Nature.* San Francisco: W. H. Freeman.

Margulis, L. 1981. *Symbiosis in Cell Evolution: Life and Its Environment on the Early Earth.* San Francisco: W. H. Freeman.

Mayr, E. 1963. *Animal Species and Evolution.* Cambridge, MA: The Belknap Press of Harvard University Press.

Monod, J. 1971. *Chance and Necessity: An Essay on the Natural Philosophy of Modern Biology.* Translated by A. Wainhouse. New York: Knopf.

Noirot, C. 1982. "La casta operaia, elemento primario del successo evolutivo delle termiti," *Rivista di Biologia/Biology Forum* 75: 157–195.

Ohno, S. 1970. *Evolution by Gene Duplication.* Berlin: Springer-Verlag.

Poppelbaum, H. 1961. *A New Zoology.* Dornach, Switzerland: Philosophic-Anthroposophic Press.

Portmann, A. 1960. *Le Forme degli Animali.* Milan: Feltrinelli.

Portmann, A. 1989. *Le Forme Viventi.* Milan: Adelphi.

Redi, F. 1668. *Experimenta circa Generationem Insectorum.* Firenze.

Reeves, R. 1963. "The Creative Art of Technology," *New Scientist* (21 July).

Rosa, D. 1932, "Teoria dell'Ologenesi." (D. Rosa, 1909–1918), alla voce "Evoluzione" della *Enciclopedia Italiana*, Vol. XIV.

Santillana, G. de, and H. von Dechend. 1983. *Il Mulino di Amleto: Saggio sul Mito e sulla Struttura del Tempo.* Milan: Adelphi.

Schrödinger, E. 1944. *What is Life? The Physical Aspect of the Living Cell.* Cambridge: Cambridge University Press.

Sermonti, G. 1984. *Mendel: Nascita e Rinascita della Genetica.* Brescia: La Scuola.

Sermonti, G., and R. Fondi. 1980. *Dopo Darwin: Critica all'Evoluzionismo.* Milan: Rusconi.

Sermonti, G., and I. Spada Sermonti. 1987. "L'ipotesi nulla nella evoluzione dei Vertebrati," *Rivista di Biologia/Biology Forum* 80: 55–77.

Sheldrake, R. 1991. *The Rebirth of Nature.* New York: Bantam Books.

Sibatani, A. 1989. "Stability of Arbitrary Structures: Its Implications for Heredity and Evolution," *Rivista di Biologia/Biology Forum* 82; 348–349.

Sibatani, A., and G. Sermonti. 1987. "Proceedings of the International Workshop on Structuralism in Biology (Osaka 1986)," *Rivista di Biologia/Biology Forum* 80 (2): 162; 269.

Spallanzani, L. 1765. *Saggio di Osservazioni Microscopiche Concernenti il Sistema della Generazione de' Signori Needham e Buffon.* Modena.

Templeton, A. R. 1983. "Phylogenetic Interference from Restriction Endonuclease Cleavage Site Maps with Particular Reference to the Evolution of Humans and the Apes," *Evolution* 37, 221–244.

Thom, R. 1975. *Structural Stability and Morphogenesis: An Outline of a General Theory of Models.* Translated by D. H. Fowler. Reading, MA: Benjamin/Cummings Publishing.

Thom, R. 1989. "Un formalismo per connettere struttura e funzione," *Rivista di Biologia/Biology Forum* 80 (3).

Thompson, D'A. W. 1992. *On Growth and Form.* New York: Dover Publications.

Tomasi Di Lampedusa, G. 1954. *Il Gattopardo.* Milan: Feltrinelli.

Vavilov, N. I. 1951. *The Origin, Variation, Immunity and Breeding of Cultivated Plants: Selected Writings.* Translated by K. S. Chester. Waltham, MA: Chronica Botanica.

Webster, G., and B. Goodwin. 1988. *Il Problema della Forma in Biologia.* Rome: Armando Editore.

Webster, G., and B. Goodwin. 1996. *Form and Transformation: Generative and Relational Principles in Biology.* Cambridge: Cambridge University Press.

Weismann, A. 1892. *Essays Upon Heredity and Kindred Biological Problems.* Translated and edited by E. B. Poulton, S. Schönland, and A. E. Shipley. Oxford: Clarendon Press.

Westenhöfer, M. 1948. *Die Grundlagen meiner Theorie vom Eigenweg des Menschen.* Heidelberg: Universitätsverlag Winter.

Wilder–Smith, A. E. 1968. *Man's Origin, Man's Destiny.* Wheaton, IL: Harold Shaw.

INDEX

Printed in the United States
86173LV00006B/174/A